U.S. Air Force Strategic Deterrence Capabilities in the 21st Century Security Environment

A Workshop Summary

Norman M. Haller, Rapporteur

Committee on U.S. Air Force Strategic Deterrence Capabilities in the 21st Century Security Environment: A Workshop

Air Force Studies Board

Division on Engineering and Physical Sciences

NATIONAL RESEARCH COUNCIL
OF THE NATIONAL ACADEMIES

THE NATIONAL ACADEMIES PRESS
Washington, D.C.
www.nap.edu

THE NATIONAL ACADEMIES PRESS   500 Fifth Street, NW   Washington, DC 20001

NOTICE: The project that is the subject of this report was approved by the Governing Board of the National Research Council, whose members are drawn from the councils of the National Academy of Sciences, the National Academy of Engineering, and the Institute of Medicine.

This is a report of work supported by Grant FA9550-12-1-0413 between the U.S. Air Force and the National Academy of Sciences. Any views expressed in this publication are those of the author(s) and do not necessarily reflect the views of the organizations or agencies that provided support for the project.

International Standard Book Number-13: 978-0-309-28547-6
International Standard Book Number-10: 0-309-28547-X

Copies of this report are available from:

The National Academies Press
500 Fifth Street, NW
Keck 360
Washington, DC 20001
(800) 624-6242 or (202) 334-3313
http://www.nap.edu

Printed in the United States of America

THE NATIONAL ACADEMIES

Advisers to the Nation on Science, Engineering, and Medicine

The **National Academy of Sciences** is a private, nonprofit, self-perpetuating society of distinguished scholars engaged in scientific and engineering research, dedicated to the furtherance of science and technology and to their use for the general welfare. Upon the authority of the charter granted to it by the Congress in 1863, the Academy has a mandate that requires it to advise the federal government on scientific and technical matters. Dr. Ralph J. Cicerone is president of the National Academy of Sciences.

The **National Academy of Engineering** was established in 1964, under the charter of the National Academy of Sciences, as a parallel organization of outstanding engineers. It is autonomous in its administration and in the selection of its members, sharing with the National Academy of Sciences the responsibility for advising the federal government. The National Academy of Engineering also sponsors engineering programs aimed at meeting national needs, encourages education and research, and recognizes the superior achievements of engineers. Dr. Charles M. Vest is president of the National Academy of Engineering.

The **Institute of Medicine** was established in 1970 by the National Academy of Sciences to secure the services of eminent members of appropriate professions in the examination of policy matters pertaining to the health of the public. The Institute acts under the responsibility given to the National Academy of Sciences by its congressional charter to be an adviser to the federal government and, upon its own initiative, to identify issues of medical care, research, and education. Dr. Harvey V. Fineberg is president of the Institute of Medicine.

The **National Research Council** was organized by the National Academy of Sciences in 1916 to associate the broad community of science and technology with the Academy's purposes of furthering knowledge and advising the federal government. Functioning in accordance with general policies determined by the Academy, the Council has become the principal operating agency of both the National Academy of Sciences and the National Academy of Engineering in providing services to the government, the public, and the scientific and engineering communities. The Council is administered jointly by both Academies and the Institute of Medicine. Dr. Ralph J. Cicerone and Dr. Charles M. Vest are chair and vice chair, respectively, of the National Research Council.

www.national-academies.org

COMMITTEE ON U.S. AIR FORCE STRATEGIC DETERRENCE CAPABILITIES IN THE 21st CENTURY SECURITY ENVIRONMENT: A WORKSHOP

GERALD F. PERRYMAN, JR., Independent Consultant, *Chair*
RAFAEL ALONSO, Science Applications International Corporation
ALLISON ASTORINO-COURTOIS, National Security Innovations, Inc.
W. PETER CHERRY, Independent Consultant
LOUIS A. COX, JR., Cox Associates, LLC
PAUL K. DAVIS, RAND Corporation
JERROLD M. POST, George Washington University
BRIAN SKYRMS, University of California at Irvine
MICHAEL O. WHEELER, Institute for Defense Analyses

Staff

CARTER W. FORD, Program Officer
SARAH M. CAPOTE, Research Associate
MARGUERITE E. SCHNEIDER, Administrative Coordinator

AIR FORCE STUDIES BOARD

Preface

Changes in the 21st century security environment require new analytic approaches to support strategic deterrence. Because current adversaries may be deterred from the use of nuclear weapons differently than were Cold War adversaries, the Air Force needs an analytic process and tools that can help determine those Air Force capabilities that will successfully deter or defeat these new nuclear-armed adversaries and assure U.S. allies. While some analytic tools are available, a coherent approach for their use in developing strategy and policy appears to be lacking. Without a coherent analytic approach that addresses the nuances of today's security environment, Air Force views of its strategic deterrence needs may not be understood or accepted by the appropriate decision makers. A coherent approach will support Air Force decisions about its strategic force priorities and needs, deter actual or potential adversaries, and assure U.S. allies.[1]

Strategic deterrence may now be far more difficult for the United States than during the Cold War. Compared to the Cold War bipolar, rational-actor model, new thinking is needed to cope with the complex notion of deterring other nuclear-armed or potentially nuclear-armed entities. As current nuclear non-peers become near-peers or peers, they may not act as expected. Non-peers that have or are developing nuclear weapons are often ruled by regimes that are difficult to penetrate, as well as regimes whose decision-making dynamics are difficult to interpret. Although these regimes may be considered irrational, other factors need to be taken into account, such as insular perspectives of adversaries; aberrant views of their role in their region; and historic, cultural, and religious biases, all of which affect the decision maker's cost-benefit calculus. U.S. security depends on having the right mix of strategic options and capabilities to deal with the new challenges. The United States may find itself engaged in a conventional war with such nuclear-armed adversaries. Some postulate that preventing escalation in such circumstances will be far more difficult than peacetime deterrence was during the Cold War. Adversaries may have powerful incentives to brandish or use nuclear weapons. It is conceivable that some nuclear-armed leaders who face very bad options may take desperate gambles, accepting a high probability of making things worse in exchange for a small hope of avoiding a large loss. Before ever getting to such a point, the Air Force must be able to understand fully and articulate convincingly its capabilities to contribute to deterrence.[2]

In this context, the Air Force in 2012 requested that the Air Force Studies Board of the National Research Council undertake a workshop to bring together national experts to discuss

---

[1]Hunter Hustus, Technical Advisor, Office of the Deputy Chief of Staff of the Air Force for Strategic Deterrence and Nuclear Integration, "USAF A10 Perspective." White paper dated September 21, 2012.

[2]Ibid.

current challenges relating strategic deterrence and potential new tools and methods that the Air Force might leverage in its strategic deterrence mission. Titled "U.S. Air Force Strategic Deterrence Capabilities in the 21st Century Security Environment," the workshop consisted of two 3-day sessions held in Washington, D.C., on September 26-28, 2012, and January 29-31, 2013.

The workshop committee was very pleased that the leaders of both Air Force organizations that championed this independent workshop, Lt Gen James Kowalski, Commander, Air Force Global Strike Command, and Maj Gen William Chambers, Assistant Chief of Staff for Strategic Deterrence and Nuclear Integration, were available to discuss in detail their needs related to this important workshop. In addition, the committee was honored that Dr. C. Paul Robinson, president emeritus of Sandia National Laboratories, former ambassador, chief U.S. negotiator, and head of the U.S. delegation to nuclear testing talks with the Soviet Union, as well as Gen Larry Welch (USAF, Ret.), trustee emeritus and former president, Institute for Defense Analyses, and former Air Force chief of staff, were able to share their perspectives in two capstone talks. Also, the committee thanks the many expert speakers and guests who contributed immensely to both sessions of this workshop.

The workshop committee's role was limited primarily to planning and organizing the workshop sessions. The workshop committee was also provided opportunities to review drafts of the workshop summary for accuracy. As a function of planning for the workshop sessions, workshop committee members exchanged e-mails and read outside materials. Some workshop committee members were asked by National Research Council staff to give presentations and moderate workshop panels as individual workshop participants.

Gerald F. Perryman, Jr., *Chair*
Committee on U.S. Air Force Strategic Deterrence Capabilities in the 21st Century Security Environment: A Workshop

Acknowledgment of Reviewers

This report has been reviewed in draft form by individuals chosen for their diverse perspectives and technical expertise, in accordance with procedures approved by the National Research Council's (NRC's) Report Review Committee. The purpose of this independent review is to provide candid and critical comments that will assist the institution in making its published report as sound as possible and to ensure that the report meets institutional standards for objectivity, evidence, and responsiveness to the study charge. The review comments and draft manuscript remain confidential to protect the integrity of the deliberative process. We wish to thank the following individuals for their review of this report:

John F. Ahearne, Sigma Xi, The Scientific Research Society,
Allison Astorino-Courtois, National Security Innovations, Inc.,
Arden L. Bement, Jr., Purdue University, and
Michael O. Wheeler, Institute for Defense Analyses.

Although the reviewers listed above have provided many constructive comments and suggestions, they were not asked to endorse the views presented at the workshop, nor did they see the final draft of the workshop summary before its release. The review of this workshop summary was overseen by Robert J. Elder, Jr., George Mason University. Appointed by the NRC, he was responsible for making certain that an independent examination of this workshop summary was carried out in accordance with institutional procedures and that all review comments were carefully considered. Responsibility for the final content of this summary rests entirely with the author and the institution.

Contents

APPENDIXES

Acronyms

AFGSC	Air Force Global Strike Command
AFGSC/CC	Commander, Air Force Global Strike Command
DoD	Department of Defense
NATO	North Atlantic Treaty Organization
NRC	National Research Council
NSC	National Security Council
SAIC	Science Applications International Corporation
TIN	timed influence net
TOR	terms of reference
USAF	U.S. Air Force
USCG	U.S. Coast Guard
USSTRATCOM	U.S. Strategic Command

1

Introduction

This report summarizes a two-part workshop titled "U.S. Air Force Strategic Deterrence Capabilities in the 21st Century Security Environment." The two workshop sessions were held in Washington, D.C., on September 26-28, 2012, and January 29-31, 2013, under the auspices of the Air Force Studies Board of the National Research Council. The workshop was attended by a very diverse set of participants with expertise in strategic deterrence and a range of analytic tools of potential interest to the Air Force. Specific terms of reference (TOR) for the workshop are listed in Box 1-1.

Early on, the workshop committee discussed the TOR, emphasizing that its work should produce something that can actually be used by the Air Force. More than once, committee members questioned whether the scope of this workshop should be limited to deterrence by "nuclear" forces or broadened to include deterrence by non-nuclear forces (e.g., conventional offensive weapons, missile defenses, cyber capabilities, space-based systems, and drones); the resulting discussion indicated that the workshop focus would be primarily on those tools and methods applicable to analysis of nuclear deterrence.[1] With respect to adjusting the TOR, the main concern was that "social network analysis and crowd sourcing" was explicitly called out, but it became clear that these terms were not meant to limit the techniques to be considered. After more discussion, the committee did not change the TOR but did develop several questions to be considered during the workshop, including the following:

1. How are the challenges for nuclear deterrence in the 21st century similar to and different from those of the 20th century?[2]
2. What are the analytic challenges, and what approaches are needed to resolve them?
3. What are the insights for the future and ancillary issues raised during workshop discussions that the Air Force should consider?

[1] Implications of cyberwarfare were not discussed extensively during the workshop.

[2] A participant noted that an additional issue was that the United States also knows more now, and if the 20th century were to be re-lived, deterrence strategy would be better. As of now, this question reflects the notion that the United States had it right in the 20th century, an interesting notion given two world wars, the Korean War, and the Vietnam War.

BOX 1-1
Terms of Reference

An ad hoc committee will plan and convene one workshop consisting of two 3-day meetings (separated for logistical reasons) to (1) examine integrated toolsets and methods, such as social network analysis and crowd sourcing, that provide insight into adversary decision calculi and insights into which Air Force capabilities are likely to be effective at influencing those decision calculi; and (2) develop terms of reference for an ad hoc study that would: (a) evaluate these integrated toolsets and insights on relevant Air Force capabilities and (b) analyze gaps.

The committee will develop the agenda for the workshop, select and invite speakers and discussants, and moderate the discussions.

In organizing the workshop, the committee might also consider additional topics close to and in line with those mentioned above. The meetings will use a mix of individual presentations, panels, breakout discussions, and question-and-answer sessions to develop an understanding of the relevant issues. Key stakeholders will be identified and invited to participate. One individually authored workshop summary document will be prepared by a designated rapporteur.[1,2]

---

[1]This workshop summary has been prepared by the workshop rapporteur as a factual summary of what occurred at the workshop. It is important to note that this rapporteur-authored workshop summary does not contain consensus findings and recommendations, which are only produced by National Research Council study committees.

[2]The terms of reference (TOR) for the workshop does not call for formal analysis and/or recommendations of how these analytic-based approaches might be used by the Air Force as part of its strategic deterrence mission; however, the notional TOR for a formal follow-on study, found in Chapter 5, does explicitly call for such analysis.

The first two questions align well with the panels and related discussions during the workshop, and the third question was explored as part of the dialog among the workshop participants at both sessions. Additionally, some speakers with a great deal of experience offered a variety of perspectives that helped establish a comprehensive backdrop for the workshop. Accordingly, the remainder of this report is organized as follows: Chapter 2, Various Perspectives; Chapter 3, Strategic Deterrence: Past, Current, and Future; Chapter 4, Analytic-Based and Non-Traditional Approaches; and Chapter 5, Insights for the Future. Finally, as a result of this workshop, the Air Force possesses a rich variety of independent thoughts regarding potential analytic approaches to substantiate Air Force concepts and articulate Air Force capabilities as deterrence strategy is developed in the 21st century security environment. The Air Force will also have illustrative elements of a TOR for a future longer-term study to evaluate potential toolsets and analyze gaps (see Chapter 5).

2

Various Perspectives

AIR FORCE PRESENTATIONS

Maj Gen William Chambers, assistant chief of staff of the Air Force for strategic deterrence and nuclear integration, Headquarters U. S. Air Force, started both workshop sessions. He observed that deterrence is not just an Air Force issue; it is a national issue. As shown in Figure 2-1, Gen Chambers emphasized that (1) the strategic deterrence challenge is different now than decades earlier, (2) for the 21st century multi-nodal world, one type of deterrence does not fit all anticipated needs, (3) ensuring stability is the preeminent goal,[1] and (4) the Air Force needs analytical tools to help it address the looming deterrence challenges.[2] These challenges, he stated, include pressures to reduce future U.S. nuclear arsenals while maintaining strategic stability as well as regional assurance in the face of actual and potential proliferation of nuclear weapons by rogue states. Gen Chambers emphasized the need to recapitalize and modernize every aspect of the nuclear deterrent, specifically calling attention to the two Air Force components of the triad (land-based missiles and bombers).[3] He also stressed a message that was revisited numerous times during the workshop—*less is different, things change as nuclear forces are reduced.* In the new strategic environment, he indicated, a new continuum of nuclear and conventional forces might become more likely. In all of this, he indicated, the Air Force needs to identify the methodologies that could provide a sound analytic basis on which to establish the Air Force strategic force requirements and priorities and justify its plans.

---

[1]During Gen Chambers' presentation, a participant posed a question about stability: "Does the other side always want stability?" This question elicited considerable discussion among the workshop participants. Gen Chambers agreed that some adversaries may actually want to foment instability.

[2]*Strategic deterrence* is described in this workshop summary as a complex-coupled problem involving many contextual factors of technical, social, political, and economic importance. These are often classified as "wicked problems" for which credible predictions (the needs for which are emphasized in several parts of this report) are unlikely. Projections based on trends, analytical insights, and measured time steps may be the best that can be expected. While barriers to predictions are not discussed in this workshop summary, the need to estimate uncertainties and error bands over time by rigorous analyses was a theme raised throughout the workshop. It may be that such analyses will require a combination of heuristics, which is also a point raised during the discussions associated with Peter Todd's presentation, "Heuristics in Uncertain Environments: Ecological Rationality," found in Chapter 2.

[3]The third element of the triad consists of submarine-based ballistic missiles.

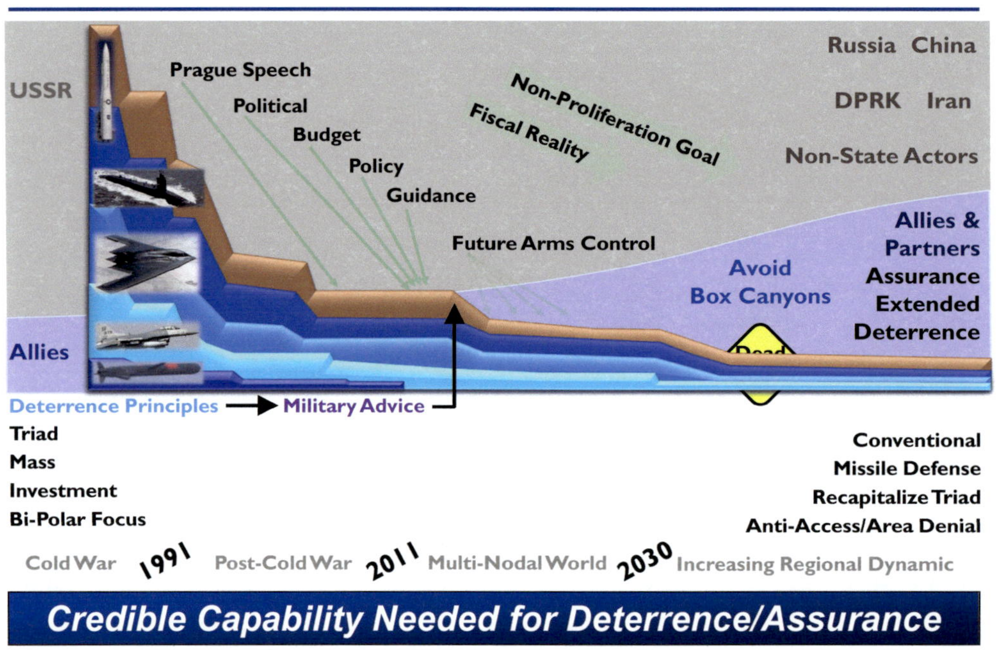

FIGURE 2-1 Ensuring that stability is the outcome. SOURCE: Maj Gen William Chambers, Assistant Chief of Staff of the Air Force for Strategic Deterrence and Nuclear Integration, "A10 Opening Remarks," presentation to the workshop on September 26, 2012.

Near the ends of both workshop sessions, Gen Chambers provided expanded discussions of his needs. A synopsis follows: Given the actors of the future, he questioned what analytical tools can be used. Current analyses, such as analyses supporting the nuclear triad, may no longer be sufficiently persuasive. He was heartened to see the intellectual capabilities being applied to this multi-disciplinary problem. Observing that regional issues are compelling, Gen Chambers noted that being able to handle the regional problem sets with analytical tools is important, so, if this workshop or any follow-on study leads to production of analyses that will help the Air Force make sound arguments for the appropriate regional flexibility, the efforts will be a success.[4] He added that it is all about investing in the right systems for the future.

Lt Gen James Kowalski, commander, Air Force Global Strike Command, emphasized in his presentation at the first workshop session that the United States is no longer in a Cold-War setting; rather, the issue is how to get from there to today and beyond, especially given the multi-polar backdrop that now exists. As shown in Figure 2-2, Gen Kowalski indicated that nuclear deterrence is the cornerstone of strategic stability (among the great powers, no one has incentive for a first strike) and underpins U.S. conventional and diplomatic power. He expressed concern about safety, security, and effectiveness of U.S. nuclear forces in light of the possibility of moving to lower numbers of nuclear weapons, for example, a few hundred deliverable warheads along with diminished capabilities provided by national laboratories and the industrial base.

[4]While not stated explicitly by Gen Chambers, an important extension of this point is the need to identify the viable alternatives for solving the problem backed up by rigorous analyses of the advantages and disadvantages of each alternative. The selection among the alternatives should be left to the decision maker.

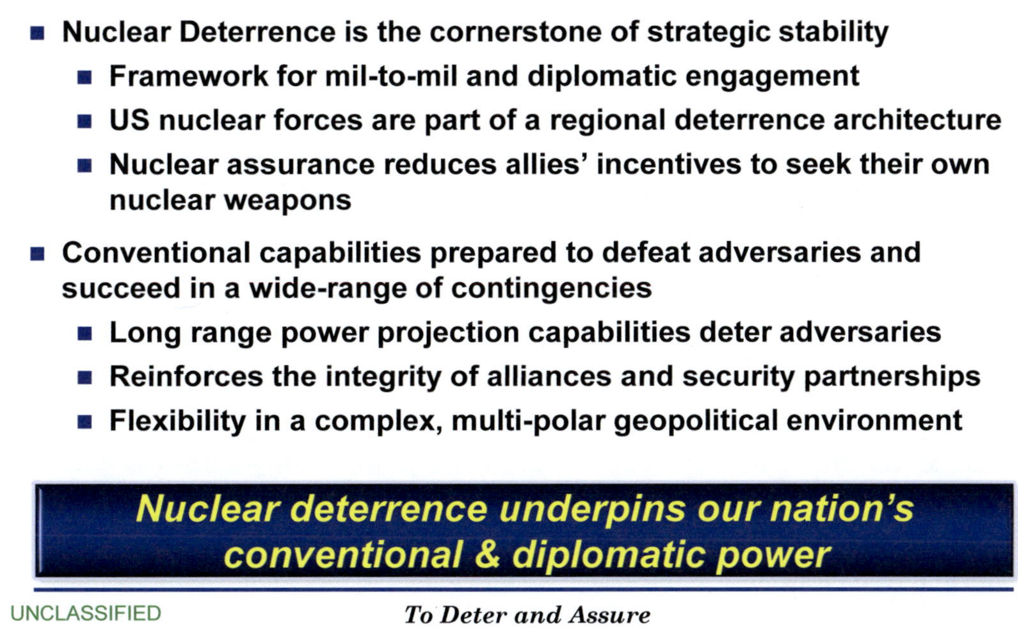

FIGURE 2-2 Air Force Global Strike Command's bottom-line mission. SOURCE: Lt Gen James Kowalski, "AFGSC Science and Technology Challenges to AFSB," presentation to the workshop on September 26, 2012.

Mr. Hunter Hustus, technical advisor, Office of the Assistant Chief of Staff of the Air Force for Strategic Deterrence and Nuclear Integration, provided a presentation to the first workshop session titled "Where Are We Now? What Is Useful?" An abstract of Mr. Hustus' presentation is found in Box 2-1.

BOX 2-1
Where Are We Now? What Is Useful?
Hunter Hustus, Office of the Assistant Chief of Staff of the Air Force for Strategic Deterrence and Nuclear Integration

The emergence of new nuclear weapons states erodes the Cold War bi-polar nature of strategic deterrence. Reductions in the size of U.S. and Russian nuclear weapons arsenals bring an end to the condition of Mutually Assured Destruction. As arsenal size decreases, the value of each warhead/system increases as does the complexity of the deterrence challenge. Policies on force structure, targeting, missile defenses, arms control agreements, and social/economic organization require new analysis. Cold War foundational constructs and analytic approaches (e.g., game theoretic) for strategic deterrence remain informative and necessary but may be insufficient for full comprehension of modern deterrence dynamics. The Air Force looks forward to the results of the workshop.

ADDITIONAL PRESENTATIONS

Dr. Daryl Press, associate professor of government, Dartmouth College, discussed future nuclear challenges at the first workshop session. The key takeaways from his presentation were (1) nuclear deterrence may be more difficult than most believe; (2) the biggest challenge is avoiding escalation during wars; and (3) there are analytic challenges regarding the future of the U.S. nuclear arsenal. Dr. Press likened today's escalation problem to the North Atlantic Treaty Organization's (NATO's) problem during the Cold War—when NATO faced what was believed to be overwhelming Soviet conventional power, and so nuclear weapons were likely to be used to stop a Soviet advance. Today the United States enjoys conventional superiority, but the roles are reversed, and less capable entities now possess nuclear weapons. In other words, adversaries may have a powerful incentive to "go nuclear," because losing a war to the United States can lead to a very bad outcome. Dr. Press closed by questioning if there is an emerging U.S. nuclear capability problem—for example, high versus low yields—and suggesting analytic-approach implications for a follow-on study to this workshop (i.e., perceptions and deterrence or capabilities and deterrence).

Lt Gen Frank Klotz (USAF, Ret.), senior fellow for strategic studies and arms control, Council on Foreign Relations, offered his insights at the second workshop session. An abstract of Gen Klotz's remarks is found in Box 2-2.

Mr. David Palkki, deputy director, Conflict Records Research Center, National Defense University, provided a presentation to the first workshop session titled "Saddam Hussein's Views on the Role of Nuclear Weapons and Perceptions Influencing his Decision-making." An abstract of Dr. Palkki's presentation is found in Box 2-3.

Dr. Peter Todd, professor of cognitive science, informatics, and psychological and brain sciences, Indiana University, provided a presentation to the second workshop session titled "Heuristics in Uncertain Environments: Ecological Rationality." An abstract of Dr. Todd's presentation is found in Box 2-4.

BOX 2-2

Achieving a Politically and Technically Sustainable Nuclear Posture for the 21st Century

Lt Gen Frank Klotz (USAF, Ret.), Council on Foreign Relations

The 2010 Nuclear Posture Review states that as long as nuclear weapons exist, the United States will sustain safe, secure, and effective nuclear forces. Other nuclear-armed states show little inclination to reduce their stockpiles to zero; some are even pursuing substantial efforts to modernize, diversify, and, in some cases, expand their existing nuclear forces. At the same time, public interest and political support for programs to maintain, much less modernize remaining U.S. nuclear capabilities have sharply declined since the end of the Cold War. Achieving consensus on the way ahead requires that two different, but not necessarily mutually exclusive beliefs be taken into account: (1) that appropriately sized nuclear forces still play an essential role in protecting U.S. and allied interests, and (2) that the United States must lead international efforts to limit and reduce nuclear arsenals, prevent proliferation, and secure nuclear materials.

BOX 2-3
Saddam Hussein's Views on the Role of Nuclear Weapons and Perceptions Influencing His Decision Making
Mr. David Palkki, National Defense University

When U.S. and U.S.-allied troops entered Iraq in 2003, they captured millions of pages of Iraqi documents and several thousand audio files of Saddam Hussein's meetings with his inner circle. These records provide unparalleled material with which to assess a recent adversary's perceptions and decision making. I present two major findings regarding Saddam's beliefs about nuclear weapons. First, Saddam and other Iraqi leaders believed that nuclear weapons provide strategic leverage, and they pursued nuclear weapons, in part, to enable conventional aggression. Iraqi acquisition of nuclear weapons would have led to violent, destabilizing Iraqi behavior. Analysts have paid too little attention to offensive, revisionist motives driving Saddam and other leaders to pursue the bomb. Second, concerns about U.S. nuclear retaliation were central to Saddam's decision not to use chemical or biological weapons in 1991. Contrary to most accounts, however, neither ambiguous U.S. nuclear threats nor U.S. threats to replace the Ba'athist regime led to Saddam's restraint.

BOX 2-4
Heuristics in Uncertain Environments: Ecological Rationality
Dr. Peter Todd, Indiana University

Traditional views of rational decision making assume that individuals should make choices by using powerful mechanisms to process all of the information available. But given that human and animal minds have evolved to be quick and just "good enough" in environments where information is often costly and difficult to obtain, we should instead expect individuals to draw on an "adaptive toolbox" of simple, fast and frugal heuristics that make good decisions with limited information and processing. These heuristics typically ignore most of the available information and rely on only a few important cues. And yet they make choices that are not only accurate when fitting their appropriate application domains, but can also be *more* accurate than traditionally rational strategies in uncertain environments—that is, when they have to generalize to new situations. Simple heuristics yield *ecological rationality* through their fit to particular information structures in the environment, and achieve their robustness in the face of environmental uncertainty via stopping rules that limit the cues they consider and so avoid overfitting noise—that is, assigning too much weight to useless cues. They also lessen the cost and other risks of gathering information. People successfully employ a variety of these heuristics in particular decision situations, such as those with time pressure and without the need to justify actions, for tasks including choosing among currently available alternatives and searching for a good-enough option out of a sequence of possibilities seen over time.

One of the discussion points after Dr. Todd's presentation concerned the availability and use of information about the types of heuristics that different leader personalities might use.

Ms. Amy Woolf, specialist in nuclear weapons policy, Congressional Research Service, shared her views on the evolution of U.S. strategic deterrence at the second workshop session. Ms. Woolf described the process of supporting Congress—a body that consists of more than 500 elected officials and thousands of staff representing interests of all the states of our nation and numerous districts within those states—a body in which most members are interested in matters other than strategic deterrence. Ms. Woolf offered that it has been useful that a small, focused group of individuals in Congress have remained interested in and committed to nuclear matters and that credible analysis could potentially be used with great effect on this group. Finally, Ms. Woolf emphasized that tightening budgets will affect congressional decisions going forward.

3

Strategic Deterrence: Past, Current, and Future

PANEL ON DETERRENCE CONCEPT UPDATES AND APPROACHES

Dr. Michael Wheeler, senior research staff, Institute for Defense Analyses, led the panel titled "Deterrence Concept Updates and Approaches" at the first workshop session. He began by describing what has and has not changed since the Cold-War (see Box 3-1). Dr. Wheeler was followed by Mr. Orde Kittre, who discussed the sanctions regime against Iran. He believes that few U.S. allies are convinced that the United States will use force against Iran over its nuclear weapons program, and he indicated that Iran's use of a nuclear weapon might be non-deterrable should it succeed in developing one. Even if Iran does not use a nuclear weapon, a nuclear-armed Iran could become emboldened. Further, he stated that several neighboring states could then also want to acquire nuclear weapons. He noted that sanctions worked in Libya, and more recent sanctions on Iran appear to be having an effect; its foreign exchange reserves are key.

Mr. Patrick McKenna, chief, Plans Evaluation and Research Division, U.S. Strategic Command (USSTRATCOM), next discussed deterrence operations and concepts of joint operations, including the official Joint Operating Concept document (which is unclassified and can be downloaded).[1] At the document's core are recognition of the changed international environment and the need to continue to deter—for example, to deter North Korea's use of a nuclear weapon in conflict. Mr. McKenna indicated that deterrence in this case means decisively influencing an adversary's decision making, which covers not only the final decision maker, but others on whom that person or small group relies in making decisions. In general, Mr. McKenna noted, the point is to deter adversary X during condition Y from doing Z, and Z could include not only nuclear-related behavior but activities in space, cyber, and proliferation, among other domains. He stated that adversary decision calculi are based on a profile. All of this is applicable to this workshop because different analytical tools might be needed for his organization's purposes—for example, understanding how an adversary might perceive use of a particular weapon, such as a high- versus low-yield nuclear weapon (e.g., bomber versus missile).

Finally, Dr. Elbridge Colby, a research analyst, provided a brief set of remarks titled "Extended Deterrence." An abstract of Dr. Colby's remarks is given in Box 3-2.

---

[1]Mr. McKenna returned for the second workshop session, at which time he discussed underlying analyses for the U.S. strategic force structure; his second presentation is summarized in Chapter 4.

BOX 3-1

Deterrence: What Has and Has Not Changed

Dr. Michael Wheeler, Institute for Defense Analyses

What Has Not Changed

1. The importance of being able to retaliate with nuclear weapons if attacked with nuclear weapons. Bernard Brodie emphasized this in his classic studies at the start of the nuclear age, as did senior Air Force leaders in the 1946 study (since unclassified) led by Generals Spaatz, Vandenberg, and Norstad.

2. Nuclear weapons are uniquely lethal and can threaten societal existence. The loss of even one city would be devastating; debates took place during the Cold War about how much damage a society could suffer before it would collapse. This was discussed in the 1950 American security review (NSC-68) led by Paul Nitze (who then was head of the Policy Planning Staff in the State Department), and in the 1950s British study by the Joint Inter-service Group for the Study of All Out War.

3. Nuclear weapons are different. A nation can lose a conventional war and recover politically, while nuclear weapons imply otherwise. North Korea has been able to threaten turning Seoul into a sea of glass for decades, but look at the intensity of diplomacy now that it has nuclear weapons. Also, look at the massive response that would be expected if a nuclear bomb ever is discovered being smuggled into a country (compared to the responses for other weapons smuggling).

4. The realities of domestic and bureaucratic politics have not changed: interagency bickering, key players being cut out, and the like. There are many examples where regional experts were excluded. For example, the Russian experts George Kennan and Chip Bohlen were kept out of the NSC-68 project (with whose conclusions they disagreed), as was Marshall Shulman (the State Department's Soviet expert during the Carter administration) during the studies leading up to Presidential Directive 59.

5. The broad outlines of the nuclear infrastructure and posture have not changed; for example, we still have three national laboratories and a triad of strategic forces.

6. Many legacies remain. For example, Russia still has the largest arsenal. Also, alliances such as NATO still rely upon the American extended deterrent.

What Has Evolved

1. Extended deterrence.

2. Proliferation challenges.

3. Arms control (e.g., the Strategic Arms Reduction Talks process, the Non-Proliferation Treaty).

What Has Changed

1. The fiscal environment and industrial base in the United States has contracted, while that in China has expanded.

2. N-party nuclear interactions are more common, as are regional interactions not directly involving the United States (as in South Asia).

Box 3-1, *continued*

3. The ubiquitous nature of the information technology revolution (database available for profiling, transparency and monitoring).

4. "Forces to the President of the United States" (nuclear, but also others: Title 10/Title 50 interactions in cyber; drone strikes; special operations raid into Pakistan to go after Osama Bin Laden).

5. No more nuclear testing/different approach to production/decline of expertise.

6. Rise of China.

Planning/Methodologies

1. U.S. Strategic Command now is the only analytic center (once had many).

2. 1960s when techniques adopted in the Department of Defense.

BOX 3-2
Extended Deterrence
Dr. Elbridge Colby, Research Analyst

Effective extended deterrence derives from a potential adversary's perception that the state extending deterrence has both the capability and the resolve to use force—possibly and perhaps necessarily including nuclear weapons—in a manner sufficiently detrimental to the potential aggressor's interests to outweigh any benefits such aggression would entail. The two key factors in effective extended deterrence are capability and resolve. Capability, in turn, can be broken down into the ability to deter through denial or through infliction of cost, with the former being more challenging. Resolve is made harder when an opponent has nuclear weapons of his own, and is especially challenging in extended deterrence because it involves the threat to use nuclear weapons for an ally's benefit by putting one's self at risk. This problem was perhaps the central one of the Cold War. While it is less central today, it remains important and may become more so. This is for two reasons: first, the United States continues to extend deterrence to over 30 countries, including a number possibly threatened by nuclear-armed adversaries; second, the U.S. conventional ascendancy of recent years appears to be narrowing; and, third, nuclear weapons appear to be proliferating to more states. The Department of Defense and the U.S. Air Force therefore need to think about what strategic deterrence capabilities are going to be required for these extended deterrence challenges.

Many workshop participants had comments and questions after the panel discussion. A synopsis follows. Mr. Kittre noted that "lawfare" is the idea that laws may be used as a tool to achieve what used to be done by military means, but there are constraints to laws (e.g., serious problems with China where the United States cannot deter cyber activity or proliferation support to other nations). There was an exchange of ideas on (1) extended deterrence (look at the costs and benefits of honoring commitments versus not honoring them) and (2) if there is an increase or decrease in an entity's caution after acquiring nuclear weapons. A workshop participant argued that there are implications to the U.S. force structure if it goes to very low numbers (e.g., to counter value instead of counter force). But Mr. McKenna indicated that strategy comes first: Would pure, city-busting force look different; and regarding timing, can it be done in 30 minutes or several weeks? However, if counter force, it likely cannot be done at lower numbers unless, perhaps, both sides go down. (A participant commented that one could also go after the other side's conventional forces.) A participant indicated that Iran is concerned that the United States wants regime change (look at what happened in Libya), and the United States tacitly accepts Pakistan and North Korean nuclear weapons, so why not a nuclear Iran eventually? A question was also raised about how one demonstrates a credible threat (e.g., B-52s, very large conventional ordnance). A participant commented that the United States is a tremendously unpredictable country, and, if provoked, it can be very decisive.

TAILORED DETERRENCE

Dr. Barry Schneider, retired director, U.S. Air Force Counterproliferation Center, provided a presentation to the first workshop session titled "Tailored Deterrence." An abstract of Dr. Schneider's presentation is found in Box 3-3.

Dr. Jerrold Post, professor of psychiatry, political psychology, and international affairs and director of the Political Psychology Program, George Washington University, provided a presentation to the first workshop session titled "Actor-Specific Behavioral Models of Adversaries: A Key Requirement for Tailored Deterrence." An abstract of Dr. Post's presentation is found in Box 3-4.

BOX 3-3
Tailored Deterrence
Dr. Barry Schneider, U.S. Air Force Counterproliferation Center (retired)

Deterrence must be tailored to (1) specific adversary leaders, (2) in specific scenarios, (3) utilizing a range of verbal and non-verbal communications, and (4) cognizant of the balance of military, economic and political power between the parties. To understand the adversary leadership, it is important to research their personality profiles, decision-making roles, propensity toward risk taking, decision processes, and their views of the U.S. leaders and credibility of U.S. deterrent threats. Where there is one dominant decision maker as there was with Saddam Hussein and Iraq, it is most important to understand that leader's personality and personal history. Where power is shared among elite, understanding and predicting is harder. However, we must try to understand how adversaries weigh costs and benefits of possible courses of action in a given set of scenarios. Further, we must discern how power is distributed within a given adversary regime, the presence of factions on different types of decisions, and their standard operating procedures, military doctrine and strategies. In addition, it is useful to know the cronies that surround top leaders and what motivates them as well as the regime's key assets and critical infrastructures and the regime's key support elements.

BOX 3-4

Actor-Specific Behavioral Models of Adversaries: A Key Requirement for Tailored Deterrence

Dr. Jerrold Post, George Washington University

One cannot extrapolate uncritically from deterrence doctrine developed during the Cold War to the post-Cold War era. Conflicts now can be precipitated by rogue leaders of outlaw nations, many of whom possess or seek to possess weapons of mass destruction. There is now no "one size fits all" in terms of deterrence, but rather the need for tailored deterrence based on actor-specific behavioral models. The profile of Saddam Hussein, offered in testimony before the House of Representatives, is presented to illustrate how a nuanced political personality profile can inform policy decisions. The profiles of three leaders of current concern are then offered: Mahmoud Ahmadinejad from Iran, the Kim Dynasty in North Korea, and Bashar al-Assad of Syria. What deters one leader may provoke another. This emphasizes the importance of an intelligence effort and analytic capabilities to develop such nuanced profiles.

CAPSTONE PRESENTATIONS

Dr. C. Paul Robinson, president emeritus, Sandia National Laboratories, provided capstone remarks at the first workshop session titled "Future Strategic Deterrence and National Security Challenges for the United States." An abstract of Dr. Robinson's remarks are found in Box 3-5. In responding to questions, Dr. Robinson provided other perspectives, such as (1) situation awareness should never be undervalued; (2) deterrence at the strategic level must rely on overwhelming fear; (3) the United States must tailor to deal with North Korea, Iran, etc., and (4) there are not enough dollars to produce the intelligence, surveillance, and reconnaissance everyone wants. Although the U.S. government must accept that there are things it will not know—it will probably know enough to communicate what is held at risk and be able to generate fear.

Gen Larry Welch (USAF, Ret.), trustee emeritus and former president, Institute for Defense Analyses, provided capstone remarks at the second workshop session titled "21st Century Deterrence." An abstract of Gen Welch's remarks are found in Box 3-6.

Gen Welch's responses to questions produced more perspectives, such as: (1) the United States should assume others are acting in what they believe are their own national interests, so it is important to understand their cultures and what their leaders believe about their true national interests; (2) the Department of Defense also needs tools to give U.S. decision makers broad understanding of what is occurring in various places; and (3) when contemplating lower levels of nuclear weapons, confidence in extended deterrence should not be lost.

BOX 3-5

Future Strategic Deterrence and National Security Challenges for the United States

Dr. C. Paul Robinson, Sandia National Laboratories (emeritus)

During the Cold War, the realization came that strategic deterrence just might be the most successful means of preventing major wars. The long peace that has extended from 1945, when nuclear weapons brought an end to the worst world war in history, continues today. The most important question for us to address is "How can we ensure that deterrence through fears of retaliation with nuclear weapons can continue ***in perpetuity*** to prevent war? This talk suggests that ***deterrence is always an active and dynamic process***, and that we must focus on the inputs to the process, if we expect the great outputs it can provide. After reviewing the history of deterrence, as seen by both Cold War protagonists, and the work carried out within the United States, one can conclude that today —with rapid changes in the world— the tasks are more complicated. We seem to be doing less well in anticipating and changing the U.S. deterrent to ensure it will remain effective for a future "multilateral nuclear-armed world." Examples discussed include: tailoring our deterrent plans for particular nations and leaders, examining changes in the target base—e.g., few if any missile fields left, more buried targets, many more mobile missiles (on underground highways?), deeply buried targets; and the characteristics of our delivery systems no longer match the targets (e.g., the low spatial density of targets obsoletes MIRVed systems, the high yields of Cold War systems no longer fit to deter less-than-major nations). The recent Air Force decision for an updated cruise missile was praised as being the likely weapon-of-choice for multilateral deterrence of less-than-major nations. The bottom line called for renewed attention to tailor the U.S. strategic deterrent to today's world.

BOX 3-6

21st Century Deterrence

Gen Larry Welch (USAF, Ret.), Institute for Defense Analyses (emeritus)

The Cold War strategic nuclear deterrence model requires expansion and adaptation to be relevant to the broader set of 21st century deterrence challenges. Still, the basic principles continue to have wide application. Further, the central Cold War nuclear deterrence task will remain relevant so long as there is the capability to destroy the United States as we know it in the hands of a government that is yet to become a reliable trustworthy friend. The most basic principle of deterrence is the need to instill in the minds of potential adversaries that the potential cost and risk of an action inimical to our interests or those of our allies far exceeds the potential gain. We were confident that we could meet that need in dealing with the leaders of the Soviet Union because we expended enormous effort over a period of decades to understand their motivations and what they valued. For deterrence to be effective on a wider scale in the 21st century, we will need to greatly increase our focus on understanding the motivations and values of a far wider and more complex set of national and trans-national actors. That understanding is essential to fashioning effective deterrent policies, strategies, and capabilities.

4

Analytic-Based and Non-Traditional Approaches

ANALYTIC-BASED APPROACHES

Dr. Paul Davis, principal researcher, Pardee Graduate School, RAND Corporation, led the panel titled "Analytic-based Approaches for Deterrence Analysis" at the first workshop session. He began by discussing some lessons learned from past work, including the following: (1) study deterrence with relatively simple models; (2) the paradigm of "rational actors" is not viable; and (3) use alternative models to defeat tyranny of best estimate since our best-estimate guesses of utility functions for the other side are not that useful and "history is replete with us getting it wrong." Dr. Davis believes much of past deterrence work has been driven by theory, and new methods can be brought to bear to inform revised theory, such as evidence from case studies, crowd-sourcing to uncover factors and mindsets, the man-machine search of data for patterns, and "historical-statistical empirical analysis." Dr. Davis added that newly developed factor-tree methods (qualitative modeling) are also quite useful in thinking about deterrence and other issues. Similarly, he noted that simple models, describable in a few viewgraphs, can frame potential adversary reasoning to help inform U.S. strategy. This can be called, with an admittedly pretentious label, synthetic cognitive modeling because it can be actor specific and highlight what the adversary worries about and has to balance.

Dr. Rob Axtell, chair, Department of Computational Social Science, Krasnow Institue for Advanced Study, George Mason University, followed with a presentation titled "Robustness and Resilience of Models Involving Social Agents." An abstract of Dr. Axtell's presentation is provided in Box 4-1.

Dr. Rita Parhad, associate partner, Monitor360, then addressed elicitation of subject-matter experts and crowd-sourcing. To illustrate the point about whether an adversary does or does not want stability, she raised a series of questions like (1) how can we tailor deterrence in that situation; (2) what can we learn from the adversary's response to our actions, messages, and policies; and (3) how can we "profile" adversaries with complex, factionalized, or opaque decision making? For deterrence to be effective she believes the socio-cultural context, along with associated motivations and actions, needs to be understood and elicitation of subject-matter experts and crowd-sourcing offer analytic techniques to gather non-U.S. perspectives and insights. Typical products of that analytical process include (1) master narratives, such as mindsets and beliefs for a country; (2) analyses of the future, such as persistent and forward-

BOX 4-1
Robustness and Resilience of Models Involving Social Agents
Dr. Rob Axtell, George Mason University

We are experiencing a revolution in the social sciences as conventional conceptions of human behavior—rationality, well-mixedness, equilibrium—are replaced by (1) the behavioral revolution in which experiments are used to elicit human behavior in specific environments, and (2) the computational revolution, in which we can scale up from the 10-25 subjects typical in the laboratory to tens of thousands or even millions of agents. Individual agent-based computational experiments in such environments can point out brittleness of policies based on optimization calculi. Systematic exploration of policy spaces can lead to more robust and resilient policies than can be predicted or achieved by other means. In order to accomplish this research program, significant resources need to be dedicated to understanding behavior in relevant domains. Specifically, regarding deterrence, signatures of sudden changes in societal behavior are not well understood, with conflicting hypotheses being advanced—e.g., some researchers claim that loss of diversity brings on rapid change while others argue that rapid growth of diversity signals abrupt transitions. An extended example from finance was indicated as a harbinger of things to come across the social sciences.

looking critical questions for a country; and (3) key influencers, such as understanding who matters in a country and how they might act, all of which could be helpful for deterrence by providing a critical context in which to make decisions.

Dr. Rafael Alonso, vice president and division manager, Autonomy and Analytics Division, Science Applications International Corporation (SAIC), finished the panel presentations by focusing on social network analysis. He indicated that most current deterrence tools do not work all that well in the context of state actors leveraging non-state actors, including terrorists. If a state is using a terrorist network, deterrence becomes very hard. He noted that it is difficult to understand the power relationships in those kinds of social networks, and the networks are seldom complete and change a lot. Mining the financial and communications data of such networks has been effective, however. He described some improvements that are underway, such as enriching text data with video or imagery and better analyzing power relationships in social networks. Dr. Alonso also touched on crowd-sourcing, suggesting its intuitive appeal due to the "safety" of large numbers may be illusory because—for example, of uninformed opinions and possible group-think—the desired result is not the same as dividing the group result by the number of individuals in the group (N).

Some key comments from various workshop participants followed. For example, all three techniques offer a framework. There is potentially useful information in these methods. After a U.S. announcement, one could sweep through responses—for example, in a country's media to suggest how the announcement was received. A participant posited sentiment mining as a tool for how to gauge what a whole population is thinking. A participant argued that these tools are more art than science. Another participant questioned how these tools could be used for deterrence. There was a suggestion that agent-based modeling captures complex behaviors, so, in principal, it can be used to see the behaviors of N states interacting. Crowd-sourcing [which refers to learning from experts and is not the same as social media or sentiment mining from a broad population], according to another participant, could be useful where leadership is centralized but cares about how the people feel (e.g., China). Another view was that for crowd-sourcing to be viable, one must get subject-matter experts engaged from day one. Crowd-sourcing is potentially useful, not just for what deters but for what can reassure allies. Several participants agreed that these tools are good for expanding knowledge of possible outcomes and allow for greater numbers of actors and interactions.

During the second workshop session, two speakers explained the analytical techniques their organizations employ for strategic forces. These presentations, summarized below, covered in more detail some of the information discussed at the first workshop session.

Mr. Patrick McKenna, chief, Plans Evaluation and Research Division, U.S. Strategic Command (USSTRATCOM), provided an overview of the analytic methodology used by USSTRATCOM to develop the desired strategic force structure and associated capabilities. He emphasized that strategy (against the backdrop of world environment) drives the force requirements. Strategy is followed by desired ends and *ways* and *means* of achieving the ends. He provided an illustrative strategic end, *deter aggression against the United States and its allies and maintain stability.* One of several effective *ways* to do that might be *demonstrate credible capability to hold at risk values or capabilities or assets an adversary values highly.* Mr. McKenna stated that one of several assessment metrics for that way could be *difference between U.S. and adversary force size in terms of prompt, survivable weapons;* this metric would then be used as part of the overall analysis of, for example, required U.S. counterforce capability (numbers and types of missiles and bombers)—the *means*. In response to a question about bringing metrics into a political debate, Mr. McKenna said his organization provides quantitative indicators with a qualitative summary, but he thought they could do better, and he challenged the workshop participants to help.

Maj Justin Sorice, scientific analyst, Air Force Office of Studies and Analyses, Assessments, and Lessons Learned, provided a presentation titled "A Framework for Strategic Deterrence Analysis." An abstract of Maj Sorice's presentation is found in Box 4-2.

BOX 4-2
A Framework for Strategic Deterrence Analysis
Maj Justin Sorice, Air Force Office of Studies and Analyses, Assessments, and Lessons Learned

An analytical framework is required to examine the interdependencies of strategies, capabilities and partnerships given planned as well as potential future reductions in the role and number of United States nuclear weapons. This presentation will examine how such a framework can be used to allow the U.S. military to think about studying strategic deterrence in the 21st century security environment, what factors and initial assumptions are required, and what conclusions are derived from the proposed framework.

NON-TRADITIONAL ANALYTIC APPROACHES

Dr. Allison Astorino-Courtois, executive vice president, National Security Innovations, Inc., set the stage by leading a panel titled “Non-Traditional Approaches to Deterrence” at the first workshop session. “How does all this relate to deterrence?” she asked rhetorically and answered, “We do not know.” She acknowledged the complex relationships, multi-actor scenarios, and nth-order effects of the 21st century and introduced a suggested set of organizing columns, as shown in Figure 4-1, which could help characterize the tools, approaches, and methods.

For example, analysis of social networks and leader profiling could be selected from a long list of possible tools (column 1) to be used to help characterize a threat (one choice from many possible purposes in column 2) in one of several domains (nuclear, non-nuclear in column 3) under conditions of peace or conflict (column 4). She noted that this organization demonstrates the broad net of conditions under which these tools might be applicable. In other words, the question must be known before the right tools can be employed.

CAPT Gail Kulisch (USCG, Ret.), Kiernan Group Holdings, provided a presentation titled “The Crafty Bastard Innovation Cycle and Solution Creation Methodology.”[1] An abstract of CAPT Kulisch’s presentation is found in Box 4-3.

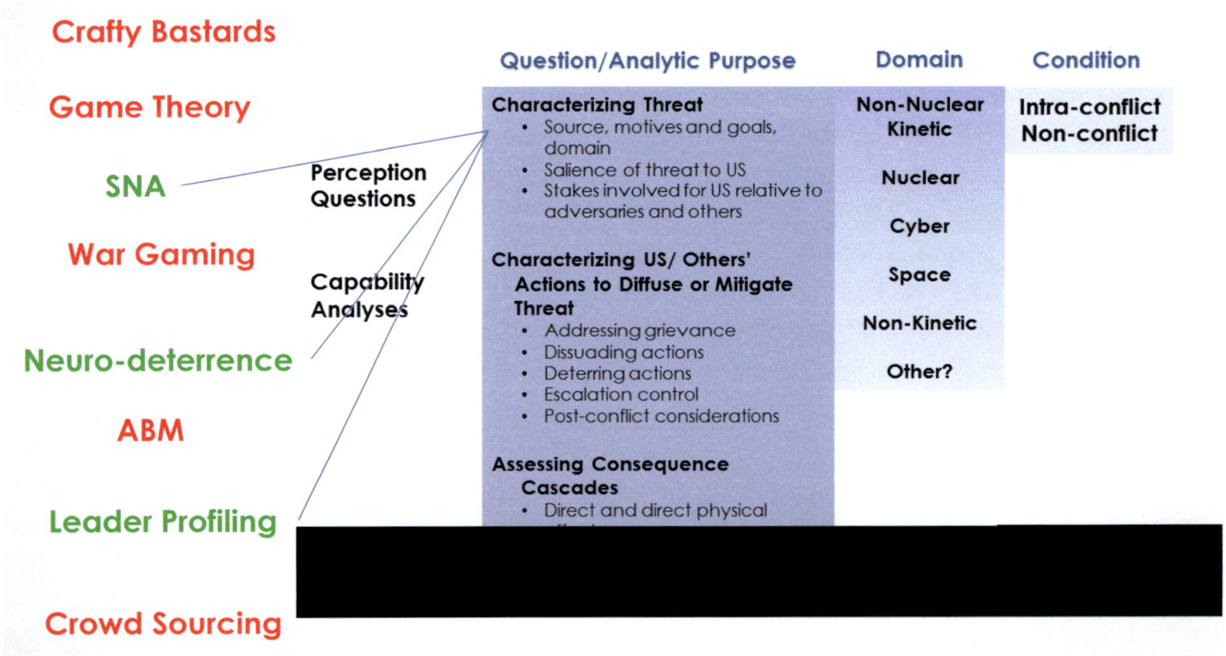

FIGURE 4-1 Characterizing tools, approaches, and methods. SOURCE: Dr. Allison Astorino-Courtois.

[1]“Crafty bastard” war gaming ("blue on red" to test alternative deterrence scenarios) recognizes the challenges of anticipating an intelligent adversary. An example would be a leader of a rogue state employing a "hard to pinpoint" subversive or terrorist organization without national or regional political responsibilities to deploy a "dirty" bomb or small nuclear device to create economic havoc (e.g., take out a major port of entry) or create fear to test national resolve and response.

Dr. Diane DiEuliis, deputy director, Office of Policy and Planning, Office of the Assistant Secretary for Preparedness and Response, U.S. Department of Health and Human Services, provided a presentation titled "Neurobiology and Deterrence: "Neurodeterrence?"An abstract of Dr. DiEuliis' presentation is found in Box 4-4.

Dr. John Sawyer, program manager/senior researcher, National Consortium for the Study of Terrorism and Responses to Terrorism, University of Maryland, described the background and current work of this center of excellence for the study of terrorism and responses to terrorism. The center advances knowledge about the causes and consequences of terrorism for homeland security policy-makers. His group within the larger organization is an incubator of innovation that embraces new methodologies and undertakes projects that may have high risk and short deadlines and be very collaborative and operationally focused. He explained one example project—essentially, a case study on ways to influence violent extremists. The methodological approach has three steps: hypothesis identification, micro literature reviews, and generation of a knowledge matrix. The resulting matrix provides access to a wide range of theories and their supporting evidence and offers a tool that could guide concepts and doctrine. Admittedly, this is a very qualitative process.

BOX 4-3

The Crafty Bastard Innovation Cycle and Solution Creation Methodology

CAPT Gail Kulisch (USCG, Ret.), Kiernan Group Holdings

The "crafty bastard" methodology is based on research and case studies which reveal that administrative, bureaucratic structures inhibit design innovation, creating self-induced constraints in a battle against unrestricted adversaries. Disruptive threats, by definition, do not fit an organization's own value chain and innovation cycle. Thinking faster than the speed of threat requires refinement of cognitive agility and the appetite for acceptance of divergent opinions and earned experience. Today's, but more importantly tomorrow's threats, demand an ability to learn more effectively and quickly and out think potential and real adversaries, whether from non-state or state sponsored organizations. Innovative and creative problem-solving design requires new hard skills that are learned through workshop encounters with exceedingly diverse and even rare combinations of talented people who work flexibly and in detail with an array of non-kinetic strategic vectors which offer innovative methods and means for unique problem sets. The crafty bastard process incorporates these critical tenets. It is an 8 week, well-defined development cycle that optimizes these attributes. Unique talent is engaged and includes experienced practitioners from the public and private sector who are practiced in applying learning and experience viewed through unique apertures that imagine new contexts. Initial analysis and assessment is conducted based on focused exploitation of deep open-source materials factoring in culture and context—the new C2 of this environment. Experienced facilitators agitate and guide critical thinking in an open, vendor-agnostic arena guided by carefully crafted questions and content. Actionable recommendations are developed and result in recommendations that disrupt the adversary's adaptation cycle, exploit emerging technologies, and generate operational and tactical level solutions that support Commander's Intent.

BOX 4-4

Neurobiology and Deterrence: "Neurodeterrence?"

Dr. Diane DiEuliis, U.S. Department of Health and Human Services

The past several decades have seen a convergence of neurobiological data with cognitive sciences, psychology and behavioral sciences—largely due to technological advances such as genomics, non-invasive imaging, and the wider availability of expansive data sets. We are thus beginning to bridge neurobiological understanding with the environmental and social backdrop upon which it occurs. An example framework for visualizing this could be the classic "OODA" loop (Observe, Orient, Decide, Act) juxtaposed onto the anatomical backdrop of limbic system function. Further affectors of this framework could be genetic predisposition, genetic heritage, and previous experience. Other fields have adapted this kind of framework for understanding the role of underlying neurobiology in human behavior and decision making, such as the emergent field of neuroeconomics. Some tenants of this field would indicate that human decision making for economics is influenced by relativistic comparisons and perceptions—and is not always rational. Similarly one could apply this framework and understanding to deterrence: understanding the underlying neurobiology that contributes to aggression, and decision making related to "intent to do harm" could provide important inputs to shaping new models and tools for the deterrence community.

Again, many questions and comments arose when the panel presentations were done. One participant asked, "If we know how people behave, does it matter what part of the brain lights up?" Dr. DiEuliis answered that this approach adds another layer of understanding. Another participant noted that, regarding the overall challenge for this workshop, an issue is how we apply neurodeterrence to what is needed for current situations, like headlines about drawing red lines related to Israel and Iran's uranium enrichment. A general issue is how can we use this. Dr. DiEuliis commented that relative to bio-weapons, in nations with lots of deaths due to diseases, bio-weapons are not that worrisome because the population is used to losing lots of people. A participant noted that some leaders are conditioned from childhood to not have empathy for others. Another participant asked whether there is any statistical way to identify some types of behavior. CAPT Kulich answered that red teams have rigor that could be helpful. In a related comment, a participant suggested that an area for research might be to observe behavior during games.

SCENARIO-BASED TOOLS

Dr. Tony Cox, president, Cox Associates, LLC, began the first panel of the second workshop session by noting that scenario-based analytic tools are meant to be used by teams, and those teams must validate both the inputs and the outputs of such tools to be sure they are trusted. A systems engineering approach in terms of an insight-generating model was presented, and analytic approaches in the context of making sense of patterns in big data were discussed. At the end, Dr. Cox explained that these presentations show the types of scenario-based tools available; much infrastructure is in place, and this represents the state of the art.

Lt Gen Robert Elder (USAF, Ret.), research professor, George Mason University, provided a presentation titled "Integrated Influence and Effects Analyses for Use in Deterrence Planning." An abstract of Gen Elder's presentation is found in Box 4-5.

Ms. Anne Russell, director, social systems analysis, SAIC, compared "old-school" analyses of data with newer approaches that augment traditional techniques with new processes, such as social network visualization and analysis techniques as well as chronological or geo-spatial visualizations like Google Earth. Among other benefits, she noted that advanced tools (e.g., a narrative pattern analyzer) can save enormous amounts of time for analysts producing outputs. An illustrative application would be assessing factionalism to help understand the degree of stability or instability of a particular country. Another illustration was use of influence-net modeling for socio-behavioral applications, which can aid reasoning under uncertainty. Ms. Russell added that the effectiveness of any one tool would depend on what the user is trying to do; it is likely that one tool would not be enough for any specific case.

BOX 4-5
Integrated Influence and Effects Analyses for Use in Deterrence Planning
Lt Gen Robert Elder (USAF, Ret.), George Mason University

Timed Influence Nets (TINs), a variant of Bayesian Nets, are used to capture cause/effect relationships that relate timed sequences of actions to the probability of an effect or outcome occurring. TIN models are thus well suited to capture the diverse aspects of nuclear strategy issues. Specifically, TIN models can be used to gain insights into the effects of actions on one or more nuclear strategy objectives and can be adapted to reflect different actors, international environments, phase of military operations, and scenarios. The TIN models can be enhanced through the use of multi-modeling techniques to leverage the ability of multi-agent-based modeling to capture the dynamic interactions among groups. TIN models were used in two service wargames and an Office of the Secretary of Defense-led geopolitical stability study to assess the deterrence and nuclear stability effects of different courses of action across a range of operational phases. The results suggest that such models can be used to inform analyses addressing nuclear policy and strategy questions.

LEADERSHIP PROFILING

Dr. Jerrold Post introduced the panel titled "Leadership Profiling Approaches" for the second workshop session, noting that leadership profiling techniques had already received much attention prior to this session. Also, he noted that trying to understand an opposing side's leadership has deep traditional roots, even though some of the approaches suggested here rightly deserve the "non-traditional" label.[2]

[2]Correctly ascertaining the intentions of an adversary is one of the key reasons why profiling was discussed so thoroughly during the workshop.

Dr. Margaret Hermann, director, Moynihan Institute of Global Affairs, Syracuse University, provided a brief set of remarks titled "Policymakers' Interpretations Matter." An abstract of Dr. Hermann's presentation is found in Box 4-6.

Dr. David Winter, Department of Psychology, University of Michigan, provided a presentation to the workshop titled "Leaders' Drives, Perceptions, and Justifications of Power: Analyzing the Signs in Crisis Situations." An abstract of Dr. Winter's presentation is provided in Box 4-7.

Dr. Stephen Walker, professor emeritus of political science, Arizona State University, provided a presentation to the workshop titled "Tailored Deterrence and Operational Code Analysis." An abstract of Dr. Walker's presentation is found in Box 4-8.

There were many comments, questions, and answers after the three presentations. For example, there was considerable discussion among participants about the pros and cons of hand-coding text versus machine coding; about getting to know the "real" persona versus the "public" persona; and the great amount of material that is available and could be analyzed. As another example, the panel speakers described what they would like to see regarding these techniques, such as (1) an easier coding process (biggest bottleneck); (2) more translations, especially of spontaneous utterances; (3) more tracing of interactions between leaders; and (4) lots of human analysis, which in the end was deemed necessary because one must look at circumstances, public statements, and what the leadership actually does (perhaps the best method). Additionally, the amount of time available to make a decision received attention from the participants—a lot of time means decentralization of power, whereas little time means "act now," which leads to contraction of power.

BOX 4-6

Policymakers' Interpretations Matter

Dr. Margaret Hermann, Syracuse University

The U.S. government employs subject matter experts to assist in the development of models to explore how particular governments are going to respond to deterrent threats and sanctions, which governments are likely to be crisis-prone, and to assess the stability of a government. What if we could, instead or in addition, determine how the leadership itself is likely to interpret a particular situation and to respond? Consider that in the past decade the 29 Asian countries located along the Pacific Rim have had 133 different governments involving changes in the leadership and their orientations to the world. The Profiler Plus software is designed to assess the leadership styles and likely behaviors of such leaders using media interviews with them, their speeches, and their written materials. The techniques have been validated by comparing results with the views of policy makers and diplomats who have interacted with the leaders.

BOX 4-7

Leaders' Drives, Perceptions, and Justifications of Power: Analyzing the Signs in Crisis Situations

Dr. David Winter, University of Michigan

In crisis situations, the intentions of the "other side" are critically important, but they are also difficult to judge. This presentation reviewed research on three concepts and measures relating to power, in order to suggest a way to estimate the aggressive intentions of potential adversaries. (1) High levels of power motive imagery in speeches, diplomatic documents, and broadcast commentaries, for example, are associated with crisis escalation. (2) In escalating crises (as compared with peacefully resolved crises), the implicit perceptions of threat each side exaggerate the threat presented by the other side, as measured by levels power imagery in summaries, précis, or "sound bites" of the other side's statements. (3) Finally, in order to secure acceptance of aggression and war by significant elites, members of the military, legislators, and ordinary citizens, leaders must frame their actions as "just," using the classical criteria suggested by Just War Theory. Taken together, these results suggest that monitoring these three measures—the other side's expression of power in political documents, the exaggerated implicit perception of threat-power in the other side's summaries of own side's statements, and the other side's justification of its power and actions—may help to estimate the intentions of potential adversaries.

BOX 4-8

Tailored Deterrence and Operational Code Analysis

Dr. Stephen Walker, Arizona State University

Tailored deterrence focuses on the problem of tailoring effective deterrence strategies to fit the beliefs, personalities, and cultural norms of diverse target populations regarding the exercise of power. The operational code construct refers to the conceptions of political strategy that inform an agent's decisions, tactics, and strategies in escalating or de-escalating conflict situations. The conceptions are measured as configurations of attributions in the public statements exchanged between agents in a strategic dyad, which index each agent's respective beliefs about the nature of the political universe (friendly or hostile), their degree of control over historical development (low or high), strategic direction (cooperation or conflict), tactical intensity (low or high), and risk-taking orientation (acceptant or averse) regarding the employment of various instruments of power (rewards, promises, threats, and punishments) in strategic interactions. These diagnostic, choice, and shift propensities are modeled formally as subjective games that each agent plays with different allies and adversaries in the political universe. The models indicate when and how members of these strategic dyads will make or respond to deterrent threats and whether such threats are necessary, desirable, or counter-productive.

THREAT ANTICIPATION AND INTELLIGENCE ANALYSIS

Dr. Michael Wheeler's opening remarks for the panel titled "Threat Anticipation and Intelligence Analysis" at the second workshop session gave an overview of the U.S. intelligence community (what it is today and how we got here) and a bit about national intelligence estimates (NIEs, the U.S. "master" estimates). He also framed how Congress got into (and stays in) the intelligence oversight business, not only through the intelligence committees but through studies it mandates, such as the one on China that was in the Fiscal Year 2013 National Defense Authorization Act. Dr. Rich Wagner, emeritus technical staff, Los Alamos National Laboratory, then provided a brief set of remarks, an abstract of which is given in Box 4-9.

BOX 4-9

How Policymakers Utilize Intelligence

Dr. Rich Wagner, Los Alamos National Laboratory

The organizing framework for thinking about U.S. strategic capabilities, including nuclear weapons, is, or should be, warning and response, over timescales ranging from minutes or hours (attack/tactical warning), to many years (strategic or geopolitical warning). The strategy should be to deliberately assess how much warning time we expect to have across this time range, and to have in place the ability to respond adequately within the warning time. This would not be just a reactive strategy; maintaining the capability to respond within warning shapes the current and future security environment. So how should the United States improve intelligence and warning capabilities to support such a strategy? (1) The powerful new capabilities for wide-area, persistent tactical intelligence, surveillance, and reconnaissance developed for Iraq, Afghanistan, and for terror interdiction can and should be adapted for nuclear attack warning and to understand other short-term threat developments short of attack. (2) The intermediate time frame of months to years is especially important for assurance of allies and for regional deterrence. Here, the model should be the NATO ShockWave program of the later phases of the Cold War. In ShockWave, full-scope U.S. and allied national intelligence, coupled with tactical ISR (which was improved expressly for these purposes), was coupled to NATO exercises designed to elicit Warsaw Pact behaviors in their subsequent exercises, in order both to help validate NATO indicator and warning capabilities and to understand changes in Pact operational concepts. (3) Nuclear weapons are mainly relevant for highest-possible-stakes geopolitical challenges of the sort that distinguished the 20th century. Since the end of the Cold War, we have been in a "strategic pause," and the question is when (if ever) and how (if at all) some highest-stakes geopolitical challenge might emerge in the future. Over the past several decades (and perhaps always), major geopolitical shifts have almost never been anticipated. "Path-gaming"—geopolitical games with notional time-scales of years or decades—have had some utility, and should be rejuvenated.

The current U.S. nuclear weapon posture is poorly suited to both near-term extended deterrence/assurance and some future major geopolitical challenge. Its main value is as a basis and starting point for its own future reconfiguration, if and when that is needed, and it should be managed expressly with that in mind.

Mr. David Hamon, principal, National and International Security Strategies, Analytic Services, Inc., provided a brief set of remarks titled "Threat Anticipation." An abstract of Mr. Hamon's remarks is found in Box 4-10.

Additional dialog among the participants surfaced other key issues, such as a need to focus on non-negotiated monitoring; a need for more effort in pulling signals out of clutter; the fact that nuclear applications need some sort of monitoring test bed; despite all the work on threat anticipation there was surprise by the "sprint-to-zero" emphasis; and—at the end of the day—humans make decisions, so there is a need to look at and understand human behavior.

BOX 4-10
Threat Anticipation
Mr. David Hamon, Analytic Services, Inc.

To Identify and develop social sciences-based research and analyses to support the anticipation and reduction of weapons of mass destruction (WMD) and related threats along a rolling long-term horizon, the Advanced Systems and Concepts Office (ASCO) of the Defense Threat Reduction Agency (DTRA), undertook a Threat Anticipation Project (TAP). Since its inception in 2002, ASCO has initiated TAP projects to explore productive areas of threat anticipation, including workshops to identify and acquire relevant expertise from the social science, computational science, and other communities; development of conceptual computer models to better understand and anticipate asymmetric threats; and staff activities involving networking with universities and other federal and private sector organizations to survey the current thinking on these issues and leverage outside expertise. TAP has produced a variety of concepts, computer and theoretical models, workshop proceedings, and reports of value to the future mission needs of DTRA. It is extremely important to have these products preserved, validated in some practical sense, and more widely used within the larger national security community. Threat anticipation by computational and social sciences is rapidly gaining recognition for potential utility; hence, TAP can become a major contributor to the Department of Defense, other federal agencies, universities, and other organizations in this area.

5

Insights for the Future

At various times during the workshop, especially at the end of the day and at the end of a session, many issues relating to strategic nuclear deterrence, the usefulness of various analytical tools, and the content of a possible follow-on consensus study were discussed by workshop participants. The sections below summarize collections of such comments by individual workshop participants, particularly those who attended the entire workshop and contributed significantly to the summary sessions. These comments reflect the considerable diversity of opinions expressed during the workshop on a range of issues. The last section contains illustrative terms of reference for a possible follow-on study.

INSIGHTS OF VARIOUS INDIVIDUAL WORKSHOP PARTICIPANTS

On Deterrence

Many participants noted that strategic deterrence, as with strategic stability, means different things to different people and that strategic deterrence is not simply the nuclear deterrence of the Cold War. The nuclear dimension, however, was the key focus of this workshop. While strategic use of conventional weapons is clearly an alternative to nuclear use, the participants did not focus on other things that could be used in a campaign. To illustrate the range of views expressed by the participants, one view was that the record shows the Russians always overestimated the United States, not so much in capability but more likely in resolve. Another view was that a rich set of challenges currently exist, such as how can the United States verify what weapons China possesses, given its extensive underground tunnels.

On General Chambers' Presentation

Gen Chambers asked, "What is it we give to the President to deter and assure? We need to develop and foster critical thinking on deterrence and assurance." These notions align with one of the Air Force's important vectors. He also reminded the participants that an examination and critical evaluation of appropriate analytic tools would be of great value to the Air Force in understanding its mission of organizing, training, and equipping two legs of the strategic triad.

On General Klotz's Presentation (Nuclear Posture)

Gen Klotz believes the administration's orientation has been clear, and there has been some consensus in Congress, but it is fragile. A participant indicated that, although the orientation is "toward" global zero, there are lots of cautions about maintaining a reliable and secure force in the meantime. A major concern also raised by this participant was that the budget tightening will lead to increased disagreements. That participant also noted that Gen Klotz partially bought into the argument that the India-Pakistan proliferation might conceivably have been avoided had the world had its act together on the Treaty on the Non-Proliferation of Nuclear Weapons and the Comprehensive Nuclear-Test-Ban Treaty, for example. On that point the participant was extremely skeptical.

A participant believed this was a very interesting presentation on the relevant issues to deterrence but noted that Gen Klotz seemed to argue for status quo with no reduction in resources, which did not seem realistic. Also, he noted that Gen Klotz did not address how to do more with less. Another participant noted that Gen Klotz's message was to not get hung up on Global Zero rhetoric; U.S. policy is not to go there unilaterally and to keep nuclear weapons safe, secure, and effective so long as others have nuclear weapons. But, that participant argued that strategy will not drive decisions made in an austere budget climate by politicians with higher priorities than nuclear deterrence. Noting that Gen Klotz believes the time is right for establishment of a new national consensus on the support and sustainment of nuclear deterrence, a participant observed that this will require two schools of thought to agree (those who say nuclear weapons are needed and those who advocate the elimination of all U.S. nuclear weapons); can both be satisfied? He concluded that this will require fact-based analysis plus the tools of such analysis.

On Congressional Perspectives

Ms. Woolf's candid presentation during the second session elicited many favorable comments, as summarized below. She indicated there tends not to be a congressional perspective, per se. Ms. Woolf noted that nuclear weapons do not have a high profile among members and that institutional knowledge has decreased over time as important members and staff members that were present during the Cold War era have retired. As a result, she stated members tend to vote along the same lines as the more knowledgeable members, which has the advantage of meaning fewer people need to be convinced. Ms. Woolf also stated that the reasoning used by staffs, often driven by advocacy groups, reflects first-order arithmetic only, which can be misleading because these calculations do not take into account underlying strategic, conceptual, or operational issues. It is possible, she argued, to change their focus, sometimes, but analysis has to be convincing and relevant to home districts or budgets.

The participants agreed that Ms. Woolf gave a fascinating description of the congressional process and explained the difficulty involved in getting traction for deterrence issues. One participant summed it up as follows: “Congress 101” means this: youth, other priorities, the Cold War took place in “ancient” times, nuclear non-proliferation and security are today's problems; what's in it for my district? cut the deficit; follow the leader, but who will lead as the experienced ones leave Congress? The participant added that the information age empowers non-governmental organizations and bloggers—a challenge for getting sound analysis into the decision process.

A participant voiced the need for showing how the maintenance of deterrence force structure at appropriate numbers can in fact be useful to a particular interest of someone who would otherwise advocate the sharp reduction or elimination of such weapons. Ms. Woolf's response was to postulate outcomes of deterrence that help satisfy something else of interest to a would-be detractor.

On Tools In General

Several participants identified many tools that may be of value to the Air Force (see Box 5-1), and many comments during the sessions related to them. A synopsis of those comments follows.

A general view by several participants was that it would be of interest to look for a suite of complementary tools and then close the aperture on bounds for possible decision making. It could be useful to do many of these, but some participants stated that they did not know how to work in some of the military environments or how it might work in a classified setting. Many participants believed they must know what information is needed and what tools could be used to get it.

Other views were as follows. One participant noted that a lot of the problem is that theory and data to support such tools is not there. Validation of such tools is most important; one would like empirical validation to be 90 percent, but there will never be an empirical way to prove some of this. Users will have to be exposed to different elements as bounding mechanisms. Another participant pointed out that there has to be some assessment of these things; how much can they be trusted? For many methods, one needs to see what works with real people.

A pessimistic view from one participant was that most of this is not ready for use now. Numbers from some of the decision tools may be worse than random. On the other hand, a more positive view came from another participant who noted that there is a huge amount of information available on new analytic techniques that is just beginning to be tapped. New concepts and methods should continue to be searched for and examined, even if some might at first be considered wild and crazy. Other participants affirmed that the real value is considering types of data that can be generated to attribute motives and perspectives to various entities (e.g., adversarial nations, terrorists). Social neuroscience research is showing "us-them" reactions and is very interesting relative to combat and ethical or moral dilemmas.

Regarding the notion of using neuroscience, a concern expressed was that one must worry about biases. A person steeped in deterrence thinking may not behave the same as a college student getting paid by the hour. Also, a lot of these studies are based on trivial tasks, which are unlikely the same as complex international tasks. One participant stated that psychological studies largely represent averages over many people, but some risk seekers and

BOX 5-1
List of Analytical Tools Considered During the Workshop

A key focus of the workshop was to identify different techniques or methods the Air Force might use in addressing strategic deterrence capabilities. At least 16 approaches were discussed:

- Qualitative analysis (international relations/strategic studies/estimative intelligence);
- Historical case studies;
- Historical statistical-empirical analysis;
- Operations research;
- Simulations and war games;
- Game theory;
- Simple deterrence analysis using synthetic cognitive models;
- Actor-specific behavioral modeling and leadership profiling;
- Agent-based computational modeling (both simple and complex cognitive decision models);
- Social network analysis/influence diagrams/data mining;
- Subject-matter-expert elicitation;
- Crowd sourcing;
- "Evil genius" and "crafty bastard" efforts;
- Insights provided by neurobiology as related to behavior;
- Heuristics; and
- Systems engineering models.

NOTE: In light of Dr. Todd's beliefs regarding the value of "simple heuristics," at least one expert cautions that with projected advances in computational capabilities, such as exaflop computing by 2020 or sooner, there will be a temptation to take a systematic modeling approach to address the higher-order complexities of deterrence techniques and capabilities.

others have different characteristics, which are not suitable for specific situations. A counter view was offered that, nevertheless, some data may be able to narrow the possibilities. Some blending of historical record and profiles with some of these techniques could have value, but one must be sure not to set decision makers up with biases.

On Profiling (Including the Panel Presentation)

Some believed strongly that there should be no shortfall on resources devoted to developing leadership profiles, which are crucial. They noted that profiling can identify tendencies, trends, and patterns, but it is not for predicting. Psychological operations are very important (for example, telling a population about luxurious life-styles of its leaders), and it is unimaginable to not know about a leader. They concluded that to augment Department of Defense decision making, more must be known about leaders. More intense intelligence effort is needed to get at closed societies.

Other participants were not convinced that profiling had significant value in all deterrence contexts, believing instead that more information is not necessarily better. Not all insights are useful; they must be tested along the way. Nevertheless, others thought profiles by different teams might help. There is merit to seeing how the other analytical tools discussed might help with profiling; social networking tools, such as sentiment mining, could play a big role in understanding adversaries and their populations. According to several participants, how to make these tools more robust is a big issue. Profiles appear to fit well with social networking tools.

The panel on leadership profiling approaches was held during the second workshop session. Drs. Winter, Walker, and Hermann mentioned being frustrated in their work by the insufficiency and immaturity of computer-based coding software plus the difficulty and time-sink of translating documents to be coded into English. A participant thought it would be better to have coding schema and tools that could handle documents in native language, but natural language coding is not there yet. Several participants believed these speakers did good work and that this is worth looking at further. They argued that the various leadership profiling approaches used in concert can yield an outcome greater than the sum of its parts.

Additional points of view were as follows. One participant stated that textual analysis—for example, use of verbs, power language, and other linguistic clues—has already been developed in some detail. These methods appear to be potentially useful for recognizing changes in leader (and influencer) attitudes and intentions based on their published speeches and remarks. A related point was made that inter-judge concordance is already high enough to suggest that these methods have some reliability. That participant also claimed that interpretation of historical experience suggests that leader language may help to predict aggressive versus less aggressive actions in escalating or resolving conflicts. Careful independent validation of these methods may be useful in determining whether they are ready for use in the context of deterrence. Extensions to detect shifts in the thinking of key influencers and shifts in power among factions, as well as hardening, softening, or changing positions or intents of factions, might be especially valuable.

Dr. Hermann and the other two panel members described approaches to leadership profiling. Although she gave fewer details than the other two, a participant thought Dr. Hermann's approach could be more amenable to computer tagging. Most thought all three panelists' work could be very useful in improving deterrence. It was noted that Dr. Winter's approach requires manual labeling of concepts from a taxonomy that includes concepts such as power imagery. He provided quantitative support for his work and noted that since it requires manual labeling it is difficult to use it with social media sources, but it could be very useful with selected document sources. A participant noted that Dr. Walker presented a very similar approach to Dr. Winter's and suggested that both approaches can be automated, but there was not a chance to discuss it further.

On Heuristics

Dr. Todd gave a presentation on an important topic. In addition to a learning tool for analysts, a participant wondered if there could be a way to help planners learn about decision biases? Another participant noted that heuristics have been shown to provide value as an aid in decision making in a variety of enterprises—but not yet in strategic deterrence. His additional views were that the key would be to pick the right heuristics and to know when and when not to rely on them; use of the wrong heuristics could be disastrous if the wrong one is picked; and application of subject-matter expertise is essential.

Participants also noted that Dr. Todd showed how, in some instances, less information is better than more. He pointed out that simpler algorithms can outperform more complex ones and provide answers in a shorter time.[1] He also described his ongoing work in cognitive bias amelioration as part of the Intelligence Advanced Research Projects Agency's *Sirius* program. Another participant observed that fast, frugal heuristics have proven useful for some problems—for example, guessing which cities are biggest based on recognition. However, they have not been studied yet in the context of deterrence (or other game theory settings, such as multi-way negotiations, or formation of a consensus decision starting from factions with different preferences). The participant added that understanding fast, frugal heuristics for conflict escalation and resolution (if they are used by people in reality) could be useful.

On Force Structure Analyses

Mr. McKenna described U.S. Strategic Command's (STRATCOM's) generic approach to analysis of force structure issues. Strategy should come first because it drives results. A participant noted that the reason for this is that they do "requirements analysis," which assesses ability to do a well-specified job, rather than characterizing capability. The participants understood that the USSTRATCOM approach handles the kinetics pretty well, but it has not done very well on issues relating to individual decision makers, political context, and world environment. USSTRATCOM does, however, consider different futures and conduct an "attribute" based parametric analysis. Regarding Mr. McKenna's framework for thinking about deterrence, the most interesting component according to some participants was his explicit separation of the overall deterrence process into ends/ways/means. He showed how to link policy/strategy to outcomes; sound analysis from two different staffs (USSTRATCOM and A9). It was also understood by the participants that sorely lacking with this type of analytical approach is an ability to understand adversary perceptions and intentions. This lack is in great contrast to the well developed ability to understand an adversary's capabilities.

On the Approach of the Air Force Office of Studies and Analyses, Assessments, and Lessons Learned

Maj Sorice described his organization's systematic efforts to analyze the implications of lower force levels across many possible conflicts with different strategies. The primary

[1]While not stated explicitly during the workshop, some experts caution that the power of computation should not extend beyond the power of comprehension. However, one should not discount the understanding that may come from modeling and simulating highly complex problems. With the expected progress over the next decade in "reverse engineering" the human brain, one can expect rapid progress in expanding "natural bridges" between what the human brain can do best with what the computer can do best.

takeaway according to many participants was a table showing that the ability to support various classic targeting options changes (or disappears) at lower force levels. The analyses were not discussed in detail; instead, Maj Sorice presented a rather detailed assessment of force postures by numerous metrics across the uncertainty space (because of concerns about classification, there were no numbers given). Participants believed his presentation offered a framework for thinking about deterrence. They noted that there were no analytic inputs (i.e., no weights associated with items), but it appeared to be a pretty complete framework for analyzing deterrence approaches. A participant observed that Maj Sorice showed how to link policy/strategy to outcomes; again, sound analysis was provided by two different staffs (USSTRATCOM and A9). The A9 organization is expecting to drive forward in fleshing out and applying the analytic framework it presented.

On Scenario-Based Analytic Tools

Gen Elder described complex modeling that is state of the art, but at least one participant questioned how it can be validated, asking, "What are the criteria for selecting subject-matter experts?" Gen Elder described an analytic framework that he developed, but at least one participant had difficulty understanding the details of its use and could not assess the value of the approach. He noted that models and integrated ensembles of models for generating insights are already available, such as Pythia, Construct, and the framework developed by the *Concepts and Analysis of Nuclear Strategy* study. A participant's view was that these models generally have uncertain validity and stop short of supporting decisions, except by providing possibly useful (but possibly misleading) insights into connections among variables. More expressive models and better validation are probably essential for closing the gap between insight and well-supported decisions.

Ms. Russell described interesting analytic tools, including Narrative Pattern Analyzer (NPA) and Influence Net Modeling (iNET/SIAM™), which show the power of new methods. A workshop participant believed, however, that it was unclear how they would be readily adapted to nuclear deterrence. Another workshop participant believed that the more useful one of the two for the purposes of this workshop appeared to be SIAM™. It provides a Bayesian framework for improving estimates with incoming information and appears capable of being employed for deterrence work with modest effort. Another participant suggested that software such as iNET and Palantir make it practical to track patterns in space and time. NPA and similar software may provide valuable clues about emerging patterns and potential threats, including shifting attitudes toward use of nuclear weapons.

On Threat Anticipation and Intelligence Analysis

Although one participant did not see any takeaways from the panel on threat anticipation and intelligence analysis, a few observations were offered. Dr. Wagner sees strategic forces as a training base for the United States to maintain knowledge and skills until—perhaps decades from now—there is again a need for "real" nuclear forces. Dr. Wagner believes we are in a strategic pause and need to be ready with appropriate analysis and analytic tools when we come out of this period. Another participant noted that Dr. Wagner did present an additional framework, the most interesting part of which was identification of a matrix that breaks up deterrence situations into four elements—negotiated monitoring/not versus treaty verification/threat.

Mr. Hamon's experience could be very valuable in developing a tool taxonomy in the future. Another participant believed there are limitations of the estimative process (manage expectations—it gets back to understanding intentions). That participant also noted that a Defense Science Board study of nuclear monitoring is expected to be available soon, and many nuclear-related studies sponsored by the DTRA's Advanced Systems and Concepts Office were done over a decade or so.

INSIGHTS FOR A FOLLOW-ON STUDY

During both workshop sessions, but especially the second, workshop participants offered many insights regarding the content of a possible follow-on study. The dialog focused on an illustrative TOR that could form a framework for such a study. Several versions of this TOR were discussed and modified during the workshop, taking into account a wide range of individual views of the participants. The notional TOR in Box 5-2 reflects comments from various participants and could serve as a starting point for decisions by the Air Force and National Academies regarding a follow-on study. During discussion of the TOR, two other suggestions offered by workshop participants were (1) for the longer-term study, why could it not look at simulation scenarios and games to see what tools might work, and (2) the study could begin with a presentation of the security environment by using a geographical schema to present conditions in applicable areas of interest; deterrence matters should be considered region by region as well as in a strategic sense.

BOX 5-2
Notional Terms of Reference for a Follow-on Study

As identified during the workshop, possible items in the terms of reference for a follow-on study by an ad hoc committee were as follows:

1. Identify the broad issues and factors that must be considered in seeking nuclear deterrence in the 21st century. Describe a program of analysis to address those issues and support planning, resourcing and managing U.S. nuclear deterrence in the 21st century.
2. Identify the major components of the analysis and the relationships among them to serve as a basis for the identification, development and use of necessary tools and methods.
3. Evaluate and recommend tools, methods, including behavioral science-based methods, and approaches for improving the understanding of how nuclear deterrence works in the 21st century, how it might fail, and how failure might be averted by the proper choice of capabilities, postures, and concepts of operation of American nuclear forces.
4. Recommend a way ahead for evolving and adapting methods and approaches in a coherent, systematic approach. This will include identifying what questions need to be addressed, and assessing what questions each tool, method, or approach is most and least valuable for this purpose.
5. Recommend how these methods and approaches can be drawn upon as a package, or used to inform each other. It is likely that any tool, method, or approach will have strengths and weaknesses.
6. Recommend criteria and a framework for validating the tools, methods, and approaches and for identifying which classes of tools, methods, and approaches are the most promising.
7. Recommend a balance of resourcing across the classes in today's austere financial climate and that can be reserved for future resourcing when and if it becomes available.

NOTE: While the workshop committee did engage in much discussion between the two workshop sessions on what could constitute the basis for the TOR of a follow-on study, the TOR reflects a much broader discussion that occurred at both workshop sessions among the many participants.

Appendix A

Biographical Sketches of Committee Members

Gerald F. Perryman, Jr. (Maj Gen, USAF, Ret.), *Chair,* is currently an independent consultant. Upon concluding military service in 2002, Gen Perryman joined Raytheon Company as vice president and lead executive for the company's Intelligence Surveillance and Reconnaissance (ISR) Strategic Business Area, McKinney, Texas. There he developed strategies for ISR growth using capabilities from across that diverse, global company, helping Raytheon to provide integrated mission systems for its many customers. From 2006 to 2011 he was director of strategic pursuits for Raytheon Intelligence and Information Systems in Garland, Texas, forming and leading teams for competitive capture of key command and control, space operations, and ISR opportunities. Prior to his Raytheon work, Gen Perryman was assistant deputy chief of staff, warfighting integration, Headquarters Air Force, providing guidance and direction for transforming Air Force warfighting capability by integrating command and control, communications and computer networks, and intelligence, surveillance, and reconnaissance systems. Earlier Gen Perryman led the Air Force's Aerospace Command and Control and ISR Center at Langley Air Force Base, Virginia. He served as commander of the 14th Air Force, which encompasses all Air Force space operations forces worldwide. Gen Perryman received his MBA from the University of North Dakota. He currently serves on the National Research Council's (NRC's) Air Force Studies Board and is a past member of the Committee on Examination of the Air Force Intelligence, Surveillance, and Reconnaissance (ISR) Capability Planning and Analysis (CP&A) Process.

Rafael Alonso is a vice president and division manager for the Autonomy and Analytics Division of Science Applications International Corporation (SAIC), where he manages a staff of more than 140 researchers and engineers in the areas of analysis, computer vision, neuroscience, robotics, remote sensing, biometrics, social media, visualization, and information systems. Dr. Alonso is also an SAIC technical fellow. Dr. Alonso joined SAIC in 2010, when SAIC acquired his previous company, SET Corporation. At SET, Dr. Alonso served as senior vice president and director of SET's Information Systems and Security Division. Prior to joining SET, Dr. Alonso was part of the management staff at Sarnoff Corporation. As technical director of Sarnoff' Convergence Laboratory, he was responsible for overseeing a staff of 40 employees with externally funded research projects in a number of areas, including multimedia storage and databases systems, web information systems, machine learning and user modeling, video quality, video compression, digital cinema, and targeted advertising. Prior to joining Sarnoff, Dr.

Alonso co-founded the Matsushita Information Technology Laboratory (MITL) in Princeton, N.J., where he served in various roles including Associate Director and Senior Scientist. At MITL, he developed leading edge information and video systems for Panasonic. Dr. Alonso started his career as an Assistant Professor in the Computer Science Department of Princeton University, where he graduated several doctoral students, and co-developed new courses in database technology and distributed systems. He has published over 50 scientific papers in information and knowledge management topics, and is currently an SAIC Fellow. Dr. Alonso obtained his B.A. in mathematics and computer science from New York University, an M.S. in electrical engineering from Columbia University, and a Ph.D. in computer science from University of California, Berkeley.

Allison Astorino-Courtois is executive vice president of National Security Innovations, Inc. (NSI), and has more than 16 years of experience in quantitative political science and decision theoretic research. Dr. Astorino-Courtois has provided lead technical management and core support for a five Department of Defense (DoD) Joint Staff and U.S. Strategic Command (USSTRATCOM) Strategic Multi-layer Analysis (SMA) projects including recently completed Competing Analysis of Nuclear Strategy for USSTRATCOM and Influencing Violent Extremist Organizations for U.S. Central Command. She has also worked a refocusing of DoD deterrence planning to the decision calculus of the actor(s) to be deterred and has designed and produced of a second- and third-order effects analysis methodology tool for military analysts and planners. Prior to joining NSI, Dr. Astorino-Courtois worked for SAIC, where among other tasks she served as a USSTRATCOM liaison to U.S. and international communities. Prior to SAIC, Dr. Astorino-Courtois was a tenured associate professor of international relations at Texas A&M University, where her research focused on the cognitive aspects of foreign policy decision making. She has received a number of academic grants and awards and has published articles in multiple peer-reviewed journals, including *International Studies Quarterly, Journal of Conflict Resolution, Political Psychology, Journal of Politics and Conflict Management*, and *Peace Science*. She has also taught at Creighton University and as a visiting instructor at the U.S. Military Academy at West Point. Dr. Astorino-Courtois earned her Ph.D. in international relations/research methodologies from New York University.

W. Peter Cherry is an independent consultant who retired in 2010 as the chief analyst on the U.S. Army's Future Combat Systems Program at SAIC. He was responsible for analytic support to requirements analysis, performance assessment, and design trades. Previously, Dr. Cherry was leader of the Integrated Simulation and Test Integrated Program Team, focusing on test and evaluation planning, the development of associated models and simulations, and the development of the Future Combat System of Systems Integration Laboratory. He was a participant in the Future Combat Systems Program from its inception, leading analysis and evaluation of concepts as a member of the Full Spectrum Team during the contract activities that preceded concept and technology development. Since the completion of his studies at the University of Michigan, Dr. Cherry has focused on the development and application of operations research in the national security domain, primarily in the field of land combat. He contributed to the development and fielding of many of the major systems employed by the Army, ranging from the Patriot Missile System to the Apache helicopter, as well as command control and intelligence systems such as ASAS and AFATDS. In addition, he contributed to the creation of the Army's Manpower Personnel and Training Program (MANPRINT) and to the Army's Embedded Training Initiative. His recent research interests include Peacekeeping Operations and the development of transformational organizations and materiel. Dr. Cherry was a member of the Army Science Board and served as chair of the Board's Logistics Subpanel. In addition he has participated over the past 10 years in independent reviews of the Army's Science and Technology programs and on NRC studies addressing a variety of defense issues. Dr. Cherry received a Ph.D. in industrial engineering from the University of Michigan. He is currently a member of the Board on Army Science and Technology, a fellow of INFORMS, and a member of the National Academy of Engineering (NAE).

Louis Anthony Cox, Jr., is president of Cox Associates, a Denver-based applied research company specializing in quantitative risk analysis, causal modeling, advanced analytics, and operations research. Since 1986, Cox Associates' mathematicians and scientists have applied computer simulation and biomathematical models, statistical and epidemiological risk analyses, causal data mining techniques, and operations research and artificial intelligence models to measurably improve health, business, and engineering risk analysis and decision making for public and private sector clients. Since 1996, its sister company, NetAdvantage, has provided operations research services and software for telecommunications companies. In 2006, Cox Associates was inducted into the Edelman Academy of the Institute for Operations Research and Management Science, recognizing outstanding real-world achievements in the practice of operations research and the management sciences. In 2012, Dr. Cox was inducted into the NAE "for applications of operations research and risk analysis to significant national problems." He has been honorary full professor of mathematics at the University of Colorado, lecturing on biomathematics, health risk modeling, computational statistics, and causality. He is on the faculties of the Center for Computational Mathematics and the Center for Computational Biology at the University of Colorado, Denver and is now a clinical professor of biostatistics and informatics at the University of Colorado Health Sciences Center. Dr. Cox holds a Ph.D. in risk analysis (1986) and an S.M. in operations research (1985), both from the Massachusetts Institute of Technology (MIT); an A.B. from Harvard University (1978); and is a graduate of the Stanford Executive Program (1993). He is a member of the NRC Board on Mathematical Sciences and Their Applications and a member of the Standing Committee on the Use of Public Health Data in FSIS Food Safety Programs.

Paul K. Davis is a senior principal researcher at the RAND Corporation and a professor of policy analysis in the Pardee RAND Graduate School. His research interests include strategic planning and methods for improving it, decision-making theory, counterterrorism, and advanced methods of analysis and modeling (notably exploratory analysis and multi-resolution modeling). He has authored or coauthored widely read books on defense planning, capabilities-based planning, portfolio analysis, and deterrence and influence theory, as well as an integrative review on social science for counterterrorism. Before joining RAND, Dr. Davis was a senior executive in DoD. He has served on numerous national panels for DoD, the National Academies, and the intelligence community. He also is a regular reviewer on several professional journals. He received his Ph.D. in chemical physics from the MIT. Dr. Davis served as a member of the

NRC Committee on Conventional Prompt Global Strike Capability and as a member of the Committee on Modeling and Simulation for Defense Transformation.

Jerrold M. Post is professor of psychiatry, political psychology, and international affairs and director of the Political Psychology Program at George Washington University. Dr. Post has devoted his entire career to the field of political psychology. Dr. Post came to George Washington after a 21-year career with the Central Intelligence Agency where he was the founding director of the Center for the Analysis of Personality and Political Behavior. He played the lead role in developing the "Camp David profiles" of Menachem Begin and Anwar Sadat for President Jimmy Carter and initiated the U.S. government program in understanding the psychology of terrorism. In recognition of his leadership at the center, Dr. Post was awarded the Intelligence Medal of Merit in 1979. He received the Nevitt Sanford Award of the International Society of Political Psychology in 2002 for Distinguished Professional Contributions to Political Psychology. In December 1990, he testified before the House Armed Services Committee and the House Foreign Affairs Committee on the political personality profile of Saddam Hussein he had developed. Since 9/11, he has testified on the psychology of terrorism before the Senate, House, and the United Nations. Dr. Post has written or edited 10 books, including *The Psychological Assessment of Political Leaders, Leaders and their Followers in a Dangerous World,* and *The Mind of the Terrorist,* and he contributed the lead chapter on "Actor-Specific Behavioral Models of Adversaries: A Key Requirement for Tailored Deterrence" in *Tailored Deterrence: Influencing States and Groups of Concern.* He is a frequent commentator in national and international media on such topics as the psychology of leadership, the psychology of terrorism, weapons of mass destruction, Osama bin Laden, Hugo Chavez, Mahmoud Ahmadinejad, Kim Jong Il, Muammar Qaddafi, and, most recently, Bashar al-Assad. Dr. Post received his baccalaureate degree magna cum laude from Yale College. After receiving his medical degree from Yale, where he was elected to Alpha Omega Alpha, honor medical society, he received post-graduate training in psychiatry at Harvard Medical School and the National Institute of Mental Health.

Brian Skyrms is a distinguished professor of logic and philosophy of science and economics at the University of California, Irvine, and a professor of philosophy at Stanford University. He has worked on problems in the philosophy of science, causation, decision theory, game theory, and the foundations of probability. Most recently, his work has focused on the evolution of social norms using evolutionary game theory. His two recent books, *Evolution of the Social Contract* and *The Stag Hunt,* are both on the topic of the workshop. These books use arguments and examples from evolutionary game theory to cover topics of interest to political philosophy, philosophy of social science, philosophy of language, and the philosophy of biology. Dr. Skyrms is a fellow of the American Academy of Arts and Sciences and one of just three living philosophers (along with Allan Gibbard and Patrick Suppes) to be elected a fellow of the National Academy of Sciences.

Michael O. Wheeler is a member of the senior research staff at the Institute for Defense Analyses (IDA), and since 1991, a member of the Strategic Advisory Group at USSTRATCOM. A 1966 graduate of the U.S. Air Force Academy, Dr. Wheeler retired in 1991 at the rank of Colonel. While in the Air Force, he served in Tactical and Strategic Air Commands, in Thailand during the Vietnam War, on the Air Staff, at the National Security Council and State Department, on the faculty of the U.S. Air Force Academy, and on the Joint Staff. At time of retirement, he was the arms control advisor to the chairman of the Joint Chiefs of Staff. In 1978-1979, Dr. Wheeler was a White House fellow. Following retirement from the Air Force, Dr. Wheeler joined strategic studies centers, first at System Planning Corporation, then at SAIC, and then at IDA. Dr. Wheeler also has served on Defense Science Board task forces and on the advisory committees for Lawrence Livermore National Laboratory and the National Nuclear Security Administration. He was the executive secretary of the congressionally chartered Commission on Nuclear Expertise (aka the Chiles Commission), and from 2006 to 2008, he was director of the Advanced Systems and Concepts Office at the Defense Threat Reduction Agency. He has published broadly in national security affairs. Dr. Wheeler holds a Ph.D. in philosophy from the University of Arizona.

Appendix B

Workshop Session Agendas

SESSION 1
SEPTEMBER 26-28, 2012
WASHINGTON, D.C.

September 26, 2012

0900 Vision for the Workshop
- Maj Gen William Chambers, Assistant Chief of Staff for Strategic Deterrence and Nuclear Integration, Headquarters U. S. Air Force, *Workshop Co-Champion*

0930 Welcome and Introductions
- Maj Gen (Ret.) Gerald Perryman, Jr., Independent Consultant

1015 Incentives for Nuclear Non-peer to Consider "First Use" of Nuclear Weapons During a Conventional Conflict with the United States or its Allies
- Dr. Daryl Press, Associate Professor, Department of Government, Dartmouth College

1115 Tailored Deterrence
- Dr. Barry Schneider, Retired Director, U.S. Air Force Counterproliferation Center

1215 Continue Discussions *with lunch available*

1300 AFGSC/CC Vision of 21st Century Deterrence
- Lt Gen James Kowalski, Commander, Air Force Global Strike Command, *Workshop Co-Champion*

1400 Saddam Hussein's Views on the Role/Utility of Nuclear Weapons and Perceptions Influencing His Decision Making
- Mr. David Palkki, Deputy Director, Conflict Records Research Center, National Defense University

1515 Where Are We Now? What Is Useful?

- Mr. Hunter Hustus, Technical Advisor, HQ USAF/A10—Strategic Deterrence and Nuclear Integration

1615 Workshop Committee Feedback to Day 1 Presentations

- All

1700 Adjourn

September 27, 2012

0900 Actor-Specific Behavioral Models of Adversaries: A Key Requirement for Tailored Deterrence

- Dr. Jerrold Post, Professor of Psychiatry, Political Psychology, and International Affairs and Director of Political Psychology Program, George Washington University

1015 Panel 1—Analytic-based Approaches for Deterrence Analysis

- Dr. Rob Axtell, Chair, Computational Social Science Department, George Mason University
- Dr. Rita Parhad, Associate Partner, Monitor360
- Dr. Rafael Alonso, Vice President and Division Manager, Autonomy and Analytics Division, Science Applications International Corporation (SAIC)

Moderator: Dr. Paul Davis, The RAND Corporation

1215 Continue Discussions *with lunch available*

1315 Panel 2—Deterrence Concept Updates and Approaches

- Dr. Elbridge Colby, Research Analyst
- Mr. Patrick McKenna, Chief, Plans Evaluation and Research Division, U.S. Strategic Command (USSTRATCOM)
- Mr. Orde Kittrie, Senior Fellow, Foundation for Defense of Democracies and Professor of Law, Sandra Day O'Connor College of Law, Arizona State University

Moderator: Dr. Michael Wheeler, Institute for Defense Analyses

1530 Workshop Committee Feedback to Day 2 Presentations

- All

1700 Adjourn

September 28, 2012

0900 Panel 3—Non-Traditional Approaches to Deterrence
- Dr. Diane DiEuliis, Deputy Director, Office of Policy and Planning, Office of the Assistant Secretary for Preparedness and Response, U.S. Department of Health and Human Services
- CAPT (Ret.) Gail Kulisch, Owner and Managing Principal of BTG Ventures, LLC
- Dr. John Sawyer, Program Manager/Senior Researcher, National Consortium for the Study of Terrorism and Responses to Terrorism, University of Maryland

Moderator: Dr. Allison Astorino-Courtois, National Security Innovations, Inc.

1100 Workshop Committee Feedback to Day 3 Presentations
- All

1200 Continue Discussions *with lunch available*

1230 Capstone: Future Strategic Nuclear Deterrence and National Security Challenges for the United States
- Dr. C. Paul Robinson, President Emeritus, Sandia National Laboratories

1300 Planning for Session 2

1400 Adjourn

SESSION 2
JANUARY 29-31, 2013
WASHINGTON, D.C.

Objectives

1. Receive briefings on topics related to the workshop terms of reference (TOR)
2. Participate in interactive panel discussions
3. Discuss potential terms of reference for follow-on National Academies' study

January 29, 2013

0900 Welcome and Introductions
- Maj Gen (Ret.) Gerald Perryman, Jr., Independent Consultant

0905 Workshop Co-Champion Opening Remarks
- Maj Gen William Chambers, Assistant Chief of Staff for Strategic Deterrence and Nuclear Integration, Headquarters U.S. Air Force

0935 An Overview of Simple Decision Heuristics in Uncertain Environments

- Dr. Peter Todd, Professor, Department of Psychological and Brain Sciences, Indiana University

1050 Panel 1—Scenario-based Analytic Tools

- Lt Gen Robert Elder (USAF, Ret.), Research Professor, George Mason University
- Ms. Anne Russell, Director of Social Systems Analysis, SAIC

Moderator: Dr. Tony Cox, Cox Associates, LLC

1200 Continue Discussions *with lunch available*

1300 Panel 2—Leadership Profiling Approaches

- Dr. David Winter, Personality and Social Contexts Chair and Professor of Psychology, University of Michigan
- Dr. Stephen Walker, Professor Emeritus of Political Sciences, Arizona State University
- Dr. Margaret Hermann, Director, Moynihan Institute of Global Affairs, Syracuse University

Moderator: Dr. Jerrold Post, George Washington University

1515 Reaction to Day 1 Presentations

- All

1700 Adjourn

January 30, 2013

0900 Achieving a Politically and Technically Sustainable Nuclear Posture for the 21st Century

- Lt Gen (Ret.) Frank Klotz, Senior Fellow for Strategic Studies and Arms Control, Council on Foreign Relations

1015 Underlying Analyses for USSTRATCOM Force Structure

- Mr. Patrick McKenna, Chief, Plans Evaluation and Research Division, USSTRATCOM

1115 Congressional Perspectives on U.S. Strategic Deterrence

- Ms. Amy Woolf, Specialist in Nuclear Weapons Policy, Congressional Research Service

1215 Recent Studies and Analyses *with lunch available*

- Maj Justin Sorice, Air Force Office of Studies and Analyses, Assessments and Lessons Learned

1315 Panel 3—Threat Anticipation and Intelligence Analysis

- Dr. David Hamon, Principal, National and International Security Strategies, Analytic Services, Inc.
- Dr. Rich Wagner, Jr., Emeritus Technical Staff, Los Alamos National Laboratory

Moderator: Dr. Mike Wheeler, Institute for Defense Analyses

1530 Reaction to Day 2 Presentations

- All

1700 Adjourn

January 31, 2013

0900 General Discussion of Potential TOR for Follow-on Consensus Study

- All

1015 Snapshot of Workshop Committee Feedback from Day 1 and Day 2

- Mr. Norm Haller, Rapporteur

1115 Capstone Remarks: Strategic Deterrence Capabilities for the 21st Century Security Environment

- Gen Larry Welch (USAF, Ret.), Trustee Emeritus and former President, Institute for Defense Analyses

1215 Continue Discussions *with lunch available*

1300 Adjourn

Appendix C

Workshop Participants

SESSION 1
SEPTEMBER 26-28, 2012
WASHINGTON, D.C.

Committee Members

Maj Gen Gerald F. Perryman, Jr. (USAF, Ret.), *Chair*
Dr. Rafael Alonso
Dr. Allison Astorino-Courtois
Dr. W. Peter Cherry (NAE)
Dr. Louis A. Cox, Jr. (NAE)
Dr. Paul K. Davis
Dr. Jerrold M. Post
Dr. Daryl G. Press
Dr. Brian Skyrms (NAS)
Dr. Michael O. Wheeler

National Research Council Staff

Mr. Terry Jaggers, *AFSB Director*
Mr. Carter Ford, *Program Officer*
Mr. Norman Haller, *Rapporteur*
Ms. Sarah Capote, *Research Associate*
Ms. Marguerite Schneider, *Administrative Coordinator*

Speakers[1]

Lt Gen James Kowalski, Commander, Air Force Global Strike Command (AFGSC), Barksdale Air Force Base, Louisiana

[1]Individual speakers were not asked to review the draft workshop summary.

Maj Gen Williams Chambers, Asistant Chief of Staff, Strategic Deterrence and Nuclear Integration, Headquarters, U.S. Air Force
Dr. Robert Axtell, Chair, Department of Computational Social Science, Center for Social Complexity, Krasnow Institute for Advanced Study, George Mason University
Dr. Elbridge Colby, Research Analyst
Dr. Diane DiEuliis, Deputy Director, Office of Policy and Planning, Office of the Assistant Secretary for Preparedness and Response, U.S. Department of Health and Human Services
Mr. Hunter Hustus, Technical Advisor, Office of the Assistant Chief of Staff of the Air Force for Strategic Deterrence and Nuclear Integration
CAPT (Ret.) Gail Kilisch, Owner and Managing Principal, BTG Ventures, LLC
Mr. Orde Kittrie, Senior Fellow, Foundation for Defense of Democracies, Professor of Law, Sandra Day O'Connor College of Law, Arizona State University
Mr. Patrick McKenna, Chief, Plans Evaluation and Research Division, USSTRATCOM
Dr. David Palkki, Deputy Director, Conflict Records Research Center, National Defense University
Dr. Rita Parhad, Associate Partner, Monitor360
Dr. C. Paul Robinson, President Emeritus, Sandia National Laboratories
Dr. Barry Schneider, Retired Director, U.S. Air Force Counterproliferation Center

Guests

Dr. James Blackwell, Special Advisor, Office of the Assistant Chief of Staff of the Air Force for Strategic Deterrence and Nuclear Integration
Mr. Scott Couture, Special Assistant to the Under Secretary, Of the Air Force for Nuclear Matters, SAF/US
Ms. Laurie Fenstermacher, Program Manager, ISR Socio-Cultural Analysis, 711th Human Performance Wing, Air Force Reserch Laboratory
Dr. Mark Gallagher, Technical Director, Air Force Office of Studies and Analyses, Assessments and Lessons Learned
Mr. Kevin Gluck, Senior Cognitive Scientist, Air Force Research Laboratory
Col. Duane Hiebsch, Chief, Strategy Division, AFGSC A8X
Maj Kevin Kippie, Chief of Nuclear Aircraft Employment, Office of the Assistant Chief of Staff of the Air Force for Strategic Deterrence and Nuclear Integration
Mr. Jeff Larsen, Senior Scientist, SAIC
Dr. Djuana Lee, Program Element Monitor, Basic Reseearch, SAF/AQRS
Mr. Darphaus Mitchell, Office of the Assistant Chief of Staff of the Air Force for Strategic Deterrence and Nuclear Integration
Dr. Edward Robbins, Supervisory Management Analyst, Office of the Assistant Chief of Staff of the Air Force for Strategic Deterrence and Nuclear Integration
Maj Kyle Smet, Nuclear Policy and Integration, Office of the Assistant Chief of Staff of the Air Force for Strategic Deterrence and Nuclear Integration
Dr. Janet Sutton, Senior Research Psychologist, 711th Human Performance Wing, Air Force Research Laboratory
Dr. John Swegle, Senior Advisory Scientist, Savannah River National Laboratory

Dr. Victor Utgoff, Senior Division Fellow, Institute for Defense Analyses
Dr. John Valentine, Chief Scientist, SAIC
Dr. Ursula Wilder, Clinical Psychologist, CIA & DNI/National Counterterrorism Center
Maj Mark Wittig, Senior Technical Advisor, USSTRATCOM
Dr. Christopher Yeaw, Chief Scientist, AFGSC

SESSION 2
JANUARY 29-31, 2013
WASHINGTON, D.C.

Committee Members

Maj Gen Gerald F. Perryman, Jr. (USAF, Ret.), *Chair*
Dr. Rafael Alonso
Dr. Allison Astorino-Courtois
Dr. W. Peter Cherry (NAE)
Dr. Louis A. Cox, Jr. (NAE)
Dr. Paul K. Davis
Dr. Jerrold M. Post
Dr. Michael O. Wheeler

National Research Council Staff

Mr. Terry Jaggers, *AFSB Director*
Mr. Carter Ford, *Program Officer*
Mr. Norman Haller, *Rapporteur*
Ms. Marguerite Schneider, *Administrative Coordinator*

Speakers

Maj Gen William A. Chambers, Assistant Chief of Staff of the Air Force for Strategic Deterrence and Nuclear Integration, Headquarters U.S. Air Force
Gen Larry D. Welch (USAF, Ret.), Trustee Emeritus and former President, Institute for Defense Analyses
Lt Gen Robert J. Elder (USAF, Ret.), Research Professor, George Mason University
Lt Gen Frank G. Klotz (USAF, Ret.), Senior Fellow for Strategic Studies and Arms Control, Council on Foreign Relations
Mr. David Hamon, Principal, National and International Security Strategies, Analytic Services, Inc.
Dr. Margaret Hermann, Director, Moynihan Institute of Global Affairs, Syracuse University
Mr. Patrick McKenna, Chief, Plans Evaluation and Research Division, USSTRATCOM
Ms. Anne Russell, Director of Social Systems Analysis, SAIC

Maj Justin E. Sorice, Scientific Analyst, Air Force Office of Studies and Analyses, Assessments and Lessons Learned
Dr. Peter M. Todd, Professor of Cognitive Science, Psychology, and Informatics, Indiana University
Dr. Rich Wagner, Jr., Emeritus Technical Staff, Los Alamos National Laboratory
Dr. Stephen Walker, Professor Emeritus of Political Science, Arizona State University
Dr. David Winter, Personality and Social Contexts Chair and Professor of Psychology, University of Michigan
Ms. Amy Woolf, Specialist in Nuclear Weapons Policy, Congressional Research Service

Guests

Dr. James Blackwell, Special Advisor, Office of the Assistant Chief of Staff of the Air Force for Strategic Deterrence and Nuclear Integration
Mr. Scott Couture, Special Assistant to the Under Secretary of the Air Force for Nuclear Matters, SAF/US
Dr. Mark Gallagher, Technical Director, Air Force Office of Studies and Analyses, Assessments and Lessons Learned
Mr. Kevin Gluck, Senior Cognitive Scientist, Air Force Research Laboratory
2d Lt Kiley Hefty, Operations Research Analyst, Air Force Office of Studies and Analyses, Assessments and Lessons Learned
Mr. Hunter Hustus, Technical Advisor, Office of the Assistant Chief of Staff of the Air Force for Strategic Deterrence and Nuclear Integration
Dr. Jeff Larsen, Senior Scientist, SAIC
Dr. Edward Robbins, Supervisory Management Analyst, Office of the Assistant Chief of Staff of the Air Force for Strategic Deterrence and Nuclear Integration
Dr. John Swegle, Senior Advisory Scientist, Savannah River National Laboratory